Reviews of Environmental Contamination and Toxicology

VOLUME 173

Springer
New York
Berlin
Heidelberg
Barcelona
Hong Kong
London
Milan
Paris
Singapore
Tokyo

Reviews of Environmental Contamination and Toxicology

Continuation of Residue Reviews

Editor
George W. Ware

Editorial Board
Lilia A. Albert, Xalapa, Veracruz, Mexico
F. Bro-Rasmussen, Lyngby, Denmark · D.G. Crosby, Davis, California, USA
Pim de Voogt, Amsterdam, The Netherlands · H. Frehse, Leverkusen-Bayerwerk, Germany
O. Hutzinger, Bayreuth, Germany · Foster L. Mayer, Gulf Breeze, Florida, USA
D.P. Morgan, Cedar Rapids, Iowa, USA · Douglas L. Park, Washington DC, USA
Raymond S.H. Yang, Fort Collins, Colorado, USA

Founding Editor
Francis A. Gunther

VOLUME 173

Springer

Coordinating Board of Editors

DR. GEORGE W. WARE, *Editor*
Reviews of Environmental Contamination and Toxicology

5794 E. Camino del Celador
Tucson, Arizona 85750, USA
(520) 299-3735 (phone and FAX)

DR. HERBERT N. NIGG, *Editor*
Bulletin of Environmental Contamination and Toxicology

University of Florida
700 Experimental Station Road
Lake Alfred, Florida 33850, USA
(941) 956-1151; FAX (941) 956-4631

DR. DANIEL R. DOERGE, *Editor*
Archives of Environmental Contamination and Toxicology

6022 Southwind Drive
N. Little Rock, Arkansas, 72118, USA
(501) 791-3555; FAX (501) 791-2499

Springer-Verlag
New York: 175 Fifth Avenue, New York, NY 10010, USA
Heidelberg: Postfach 10 52 80, 69042 Heidelberg, Germany

Library of Congress Catalog Card Number 62-18595.
Printed in the United States of America.

ISSN 0179-5953

Printed on acid-free paper.

© 2002 by Springer-Verlag New York, Inc.
All rights reserved. This work may not be translated or copied in whole or in part without the written permission of the publisher (Springer-Verlag New York, Inc., 175 Fifth Avenue, New York, NY 10010, USA), except for brief excerpts in connection with reviews or scholarly analysis. Use in connection with any form of information storage and retrieval, electronic adaptation, computer software, or by similar or dissimilar methodology now known or hereafter developed is forbidden. The use of general descriptive names, trade names, trademarks, etc., in this publication, even if the former are not especially identified, is not to be taken as a sign that such names, as understood by the Trade Marks and Merchandise Marks Act, may accordingly be used freely by anyone.

ISBN 0-387-95339-6   SPIN 10847030

Springer-Verlag   New York Berlin Heidelberg
A member of BertelsmannSpringer Science+Business Media GmbH

Foreword

International concern in scientific, industrial, and governmental communities over traces of xenobiotics in foods and in both abiotic and biotic environments has justified the present triumvirate of specialized publications in this field: comprehensive reviews, rapidly published research papers and progress reports, and archival documentations. These three international publications are integrated and scheduled to provide the coherency essential for nonduplicative and current progress in a field as dynamic and complex as environmental contamination and toxicology. This series is reserved exclusively for the diversified literature on "toxic" chemicals in our food, our feeds, our homes, recreational and working surroundings, our domestic animals, our wildlife and ourselves. Tremendous efforts worldwide have been mobilized to evaluate the nature, presence, magnitude, fate, and toxicology of the chemicals loosed upon the earth. Among the sequelae of this broad new emphasis is an undeniable need for an articulated set of authoritative publications, where one can find the latest important world literature produced by these emerging areas of science together with documentation of pertinent ancillary legislation.

Research directors and legislative or administrative advisers do not have the time to scan the escalating number of technical publications that may contain articles important to current responsibility. Rather, these individuals need the background provided by detailed reviews and the assurance that the latest information is made available to them, all with minimal literature searching. Similarly, the scientist assigned or attracted to a new problem is required to glean all literature pertinent to the task, to publish new developments or important new experimental details quickly, to inform others of findings that might alter their own efforts, and eventually to publish all his/her supporting data and conclusions for archival purposes.

In the fields of environmental contamination and toxicology, the sum of these concerns and responsibilities is decisively addressed by the uniform, encompassing, and timely publication format of the Springer-Verlag (Heidelberg and New York) triumvirate:

Reviews of Environmental Contamination and Toxicology [Vol. 1 through 97 (1962–1986) as Residue Reviews] for detailed review articles concerned with any aspects of chemical contaminants, including pesticides, in the total environment with toxicological considerations and consequences.

Bulletin of Environmental Contamination and Toxicology (Vol. 1 in 1966) for rapid publication of short reports of significant advances and discoveries in the fields of air, soil, water, and food contamination and pollution as well as methodology and other disciplines concerned with the introduction, presence, and effects of toxicants in the total environment.

Archives of Environmental Contamination and Toxicology (Vol.1 in 1973) for important complete articles emphasizing and describing original experimental or theoretical research work pertaining to the scientific aspects of chemical contaminants in the environment.

Manuscripts for *Reviews* and the *Archives* are in identical formats and are peer reviewed by scientists in the field for adequacy and value; manuscripts for the *Bulletin* are also reviewed, but are published by photo-offset from camera-ready copy to provide the latest results with minimum delay. The individual editors of these three publications comprise the joint Coordinating Board of Editors with referral within the Board of manuscripts submitted to one publication but deemed by major emphasis or length more suitable for one of the others.

Coordinating Board of Editors

Preface

Thanks to our news media, today's lay person may be familiar with such environmental topics as ozone depletion, global warming, greenhouse effect, nuclear and toxic waste disposal, massive marine oil spills, acid rain resulting from atmospheric SO_2 and NO_x, contamination of the marine commons, deforestation, radioactive leaks from nuclear power generators, free chlorine and CFC (chlorofluorocarbon) effects on the ozone layer, mad cow disease, pesticide residues in foods, green chemistry or green technology, volatile organic compounds (VOCs), hormone- or endocrine-disrupting chemicals, declining sperm counts, and immune system suppression by pesticides, just to cite a few. Some of the more current, and perhaps less familiar, additions include *xenobiotic transport, solute transport, Tiers 1 and 2, USEPA to cabinet status, and zero-discharge*. These are only the most prevalent topics of national interest. In more localized settings, residents are faced with leaking underground fuel tanks, movement of nitrates and industrial solvents into groundwater, air pollution and "stay-indoors" alerts in our major cities, radon seepage into homes, poor indoor air quality, chemical spills from overturned railroad tank cars, suspected health effects from living near high-voltage transmission lines, and food contamination by "flesh-eating" bacteria and other fungal or bacterial toxins.

It should then come as no surprise that the '90s generation is the first of mankind to have become afflicted with *chemophobia*, the pervasive and acute fear of chemicals.

There is abundant evidence, however, that virtually all organic chemicals are degraded or dissipated in our not-so-fragile environment, despite efforts by environmental ethicists and the media to persuade us otherwise. However, for most scientists involved in environmental contaminant reduction, there is indeed room for improvement in all spheres.

Environmentalism is the newest global political force, resulting in the emergence of multi-national consortia to control pollution and the evolution of the environmental ethic. Will the new politics of the 21st century be a consortium of technologists and environmentalists or a progressive confrontation? These matters are of genuine concern to governmental agencies and legislative bodies around the world, for many serious chemical incidents have resulted from accidents and improper use.

For those who make the decisions about how our planet is managed, there is an ongoing need for continual surveillance and intelligent controls to avoid endangering the environment, the public health, and wildlife. Ensuring safety-in-use of the many chemicals involved in our highly industrialized culture is a dynamic challenge, for the old, established materials are continually being displaced by newly developed molecules more acceptable to federal and state regulatory agencies, public health officials, and environmentalists.

Adequate safety-in-use evaluations of all chemicals persistent in our air, foodstuffs, and drinking water are not simple matters, and they incorporate the judgments of many individuals highly trained in a variety of complex biological, chemical, food technological, medical, pharmacological, and toxicological disciplines.

Reviews of Environmental Contamination and Toxicology continues to serve as an integrating factor both in focusing attention on those matters requiring further study and in collating for variously trained readers current knowledge in specific important areas involved with chemical contaminants in the total environment. Previous volumes of *Reviews* illustrate these objectives.

Because manuscripts are published in the order in which they are received in final form, it may seem that some important aspects of analytical chemistry, bioaccumulation, biochemistry, human and animal medicine, legislation, pharmacology, physiology, regulation, and toxicology have been neglected at times. However, these apparent omissions are recognized, and pertinent manuscripts are in preparation. The field is so very large and the interests in it are so varied that the Editor and the Editorial Board earnestly solicit authors and suggestions of underrepresented topics to make this international book series yet more useful and worthwhile.

Reviews of Environmental Contamination and Toxicology attempts to provide concise, critical reviews of timely advances, philosophy, and significant areas of accomplished or needed endeavor in the total field of xenobiotics in any segment of the environment, as well as toxicological implications. These reviews can be either general or specific, but properly they may lie in the domains of analytical chemistry and its methodology, biochemistry, human and animal medicine, legislation, pharmacology, physiology, regulation, and toxicology. Certain affairs in food technology concerned specifically with pesticide and other food-additive problems are also appropriate subjects.

Justification for the preparation of any review for this book series is that it deals with some aspect of the many real problems arising from the presence of any foreign chemical in our surroundings. Thus, manuscripts may encompass case studies from any country. Added plant or animal pest-control chemicals or their metabolites that may persist into food and animal feeds are within this scope. Food additives (substances deliberately added to foods for flavor, odor, appearance, and preservation, as well as those inadvertently added during manufacture, packing, distribution, and storage) are also considered suitable review material. Additionally, chemical contamination in any manner of air, water, soil, or plant or animal life is within these objectives and their purview.

Normally, manuscripts are contributed by invitation, but suggested topics are welcome. Preliminary communication with the Editor is recommended before volunteered review manuscripts are submitted.

Tucson, Arizona   G.W.W.

Table of Contents

Foreword v
Preface vii

Association Between Contaminant Tissue Residues and Effects in
Aquatic Organisms 1
    MACE G. BARRON, JAMES A. HANSEN, AND JOSHUA LIPTON

Toxicity of Azaarenes 39
    ERIC A.J. BLEEKER, SASKIA WIEGMAN, PIM DE VOOGT,
    MICHIEL KRAAK, HEATHER A. LESLIE, ELSKE DE HAAS,
    AND WIM ADMIRAAL

Enantiomeric Enrichment of Chiral Pesticides in the Environment 85
    WIM J.M. HEGEMAN AND REMI W.P.M. LAANE

RMS *Titanic* and the Emergence of New Concepts on Consortial Nature
of Microbial Events 117
    D. ROY CULLIMORE, CHARLES PELLEGRINO, AND LORI JOHNSTON

Index 143

Information for Authors 147

Association Between Contaminant Tissue Residues and Effects in Aquatic Organisms

Mace G. Barron, James A. Hansen, and Joshua Lipton

Contents

I. Introduction	1
II. The Critical Body Residue Approach	2
A. Theoretical and Experimental Basis of CBRs	2
B. Uncertainties in the CBR Approach	6
III. Methodology for Evaluating CBRs	6
IV. Narcotic Chemicals	7
A. Background	7
B. Variability of CBRs	9
C. Determinants of CBRs	10
V. Nonnarcotic Organic Chemicals	14
A. Excitatory Agents	14
B. Acetylcholinesterase Inhibitors	17
C. Reactive Chemicals	18
D. CNS Seizure Agents	19
E. Aryl Hydrocarbon Receptor Agonists	20
VI. Metals	23
A. Inorganic Metals	23
B. Organometallic Chemicals	25
VII. Discussion	26
Summary	29
References	31

I. Introduction

Tissue residues have been proposed to be a more appropriate indicator of adverse effects in aquatic biota than external water concentrations because tissue residues should represent a more toxicologically relevant "dose" (McCarty and Mackay 1993). McCarty (1986, 1987) examined quantitative structure—activity relationships (QSARs) of the acute toxicity and bioconcentration of organic chemicals and concluded that chemicals should accumulate to a critical body residue (CBR). McCarty (1986) defined the CBR as the molar tissue concentra-

---

Communicating Editor: G.W. Ware.

M.G. Barron (✉)
P.E.A.K. Research, 1134 Avon Lane, Longmont, CO 80501, U.S.A.

J.A. Hansen·J. Lipton
Stratus Consulting, 1881 9th Street, Suite 201, Boulder, CO 80306, U.S.A.

tion (e.g., mmol/kg) of a toxic chemical that consistently produces a defined toxic effect such as mortality or reduced growth. According to the theory of McCarty and others (e.g., McCarty and Mackay 1993), CBRs within a defined mode-of-action category should be relatively constant across different chemicals, species, and exposure conditions. Although the CBR concept has both theoretical and experimental support, a comprehensive evaluation of the consistency and applicability of the CBR approach for different chemical classes has not previously been completed.

This review evaluates the consistency of CBRs and the applicability of the CBR approach for eight mode-of-action and chemical classes. These mode-of-action classes encompassed a broad range of environmental contaminants and included narcotics, excitatory agents, acetylcholinesterase (AChE) inhibitors, reactives/irritants, central nervous system (CNS) seizure agents, aryl hydrocarbon (Ah) receptor agonists, inorganic metals, and organometals. The organic chemicals within these classes have been previously defined based on evaluations of chemical structure, degree of toxicity, and behavioral effects in fish (Russom et al. 1997). The primary objectives of this review are (1) to determine if representative chemicals within a defined mode-of-action class accumulate to a consistent "critical" level in different species of aquatic organisms and (2) to determine if CBRs within each mode-of-action class are relatively consistent for a variety of chemicals and exposure conditions.

This review is organized as follows. Section II provides an overview of the CBR approach and Section III presents the methodology used to evaluate CBRs in this review. Section IV discusses data related to CBRs for narcotics. This chemical class is treated separately because of the substantial amount of attention devoted to developing CBRs for chemicals with a narcosis mode of action (McCarty 1986, 1987; McCarty et al. 1993). Section V considers and evaluates tissue residue data for other classes of organic chemicals: excitatory agents, AChE inhibitors, reactive compounds, CNS seizure agents, and compounds whose toxicity has been associated with binding to the Ah receptor. Section VI discusses the more limited information available regarding metals, including both heavy metals and organometallic compounds. Section VII discusses the physical, chemical, and biological determinants of CBRs and provides conclusions and research recommendations.

II. The Critical Body Residue Approach
A. Theoretical and Experimental Basis for CBRs

A CBR is defined as the concentration of a chemical accumulated in tissues of an aquatic organism that corresponds to a specific toxicity endpoint such as mortality, reduced growth, or reduced reproduction. Specifically, a tissue residue concentration is considered a "critical" residue if it is consistently associated with a specific toxicity endpoint, independent of aqueous exposure conditions. The body residue for a specific chemical therefore cannot be considered a CBR if exposure conditions or other chemical, biological, or environmental variables substantially modify the body residue concentration associated with the specific toxicity endpoint. A CBR can be defined for either individual chemicals or for classes of chemicals that share the same mode of action, and have been hypothesized to be relatively constant across a wide range of aquatic species and taxonomic groups (McCarty and Mackay 1993).

Evaluation of chemical toxicity using tissue residues rather than an evaluation of toxicity based on contaminant concentrations in water, sediment, or diet offers several potential advantages. These advantages may include providing a direct measure of the internally accumulated dose, an indication of site-specific bioavailability, and integration of contaminant exposure routes and duration. In contrast, chemical toxicity based on the aqueous exposure concentration has been shown to be modified by the duration of the exposure, the history of an organism's prior exposure, the exposure dynamics (e.g., intermittent, continuous, or pulsed exposure concentrations), the bioavailability of the chemical, and the route of exposure. These chemical, biological, and environmental variables produce toxicity values for aqueous exposures of narcotic chemicals that encompass a range of values differing by four to five orders of magnitude (McCarty 1987).

The relationship between toxicological effects and exposure can be described as a CBR if tissue residue concentrations consistently produce the same toxicological effect (Fig. 1). Figure 1a shows that both water concentration and tissue residue are associated with adverse effects. In this case, the tissue concentration may or may not causally determine effects, depending on whether the tissue residue represents the toxicologically relevant dose or whether it simply covaries with the water exposure. In Fig. 1b, tissue residue, but not water concentration, is associated with adverse effects and therefore is potentially "critical." Figure 1c illustrates the reverse situation: water concentration is associated with effects, whereas tissue residues are not. Figure 1d illustrates the CBR concept by showing that the association between water concentration and adverse effects varies with exposure conditions, pH in this illustration, whereas tissue residue is a consistent predictor of a specific adverse effect level.

The CBR concept was developed by McCarty and colleagues (McCarty 1986, 1987; McCarty et al. 1992, 1993) after they determined that tissue concentrations of many chemicals with the same mode of toxic action were relatively constant for a defined level of toxicity such as death, when expressed on a molar basis (mmol/kg). McCarty (1987) predicted that CBRs for narcotic chemicals would be 1–2 mmol/kg for acute toxicity and 0.2–0.4 mmol/kg for chronic toxicity. CBRs for chemicals with more specific modes of action such as receptor-mediated toxicity were predicted to be lower than those for narcotic chemicals. For example, McCarty (1987) predicted the CBRs for nonnarcotic chemicals to be 0.3– 0.6 mmol/kg for acute toxicity and 0.03–0.2 mmol/kg for chronic toxicity.

McCarty et al. (1992, 1993) later refined the CBR estimates for narcotic chemicals into two mode-of-action subclasses: nonpolar narcotics, which do not contain an oxygen or other polar group, and polar narcotics, which contain an

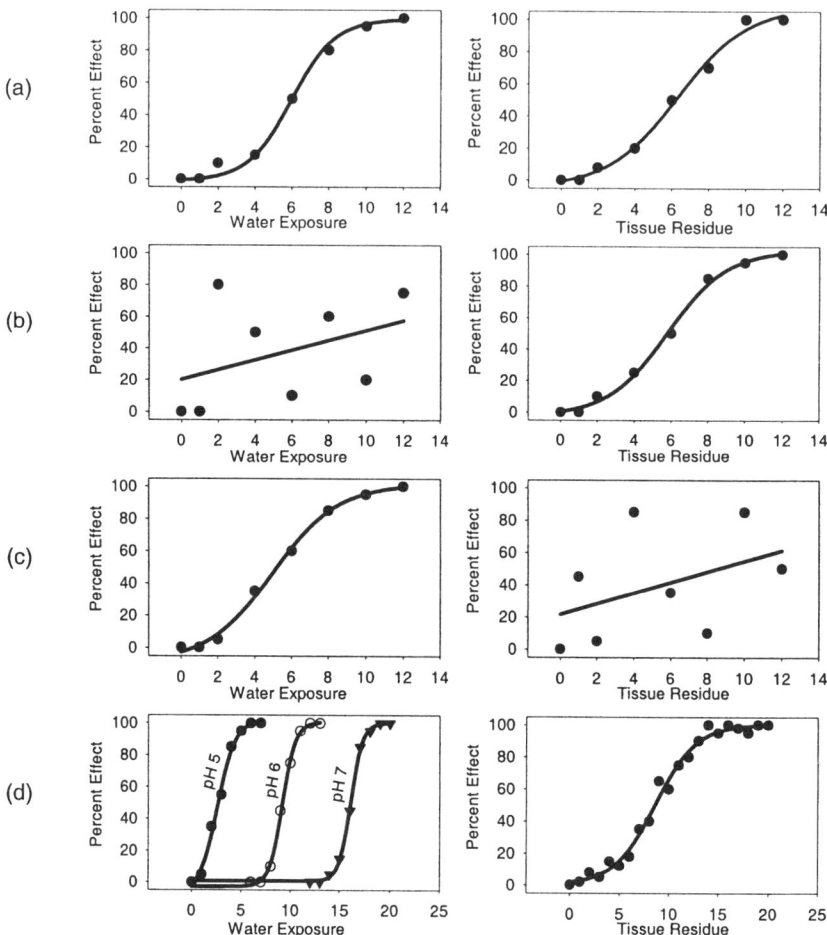

Fig. 1. Hypothetical relationships between exposure and toxicological effects in which exposure is represented as either water or tissue concentration. a, b. Potential critical body residue (CBR) relationships between dose and response. c. Relationship between dose and response that does not support the CBR concept. d. Hypothetical tissue dose–response relationship that supports the CBR concept.

ionizable polar group such as a hydroxyl. Nonpolar narcotics were predicted to have a CBR for acute lethality of 4.4 mmol/kg. The predicted CBR was supported by the relatively narrow range (>0.7–13 mmol/kg) of experimentally determined and estimated CBRs for several chlorobenzenes, polychlorinated biphenyls (PCBs), polycyclic aromatic hydrocarbons (PAHs), and the carbamate insecticide aminocarb (McCarty et al. 1992). McKim and Schmieder (1991) calculated lethal residues of 14 organic chemicals in large rainbow trout (*Oncorhynchus mykiss*) from chemical uptake experiments and grouped the CBR data according to the mode-of-action characteristics of the chemical. In agreement with the CBR concept, the two nonpolar narcotics had the highest calculated CBRs (1.6 and 1.7 mmol/kg) and the polar narcotics had the lowest calculated CBRs (0.2–1 mmol/kg).

McCarty et al. (1993) estimated CBRs for 30 phenolic compounds in fathead minnows (*Pimephales promelas*) from the relationship between acute toxicity (LC_{50}) and bioconcentration factors (BCFs). The estimated CBRs encompassed a 1000-fold range in values, which was attributed to differences in the mode of action of the compounds. Less polar alkyl phenols generally accumulated to the highest CBRs (>2 mmol/kg) and were considered by McCarty et al. (1993) to be nonpolar narcotics, whereas those with CBR values between 0.5 and 2.0 mmol/kg were considered to be polar narcotics. Phenolic compounds with CBRs less than 0.5 mmol/kg were presumed to have a nonnarcotic mode of action. McCarty et al. (1993) concluded that the CBR values estimated for fathead minnows were in general agreement with experimentally determined and theoretically estimated CBR data for other aquatic species. However, they also concluded that there was substantial uncertainty in the CBRs because of uncertainty in the type and number of modes of action of the phenolic compounds examined.

McCarty and Mackay (1993) summarized the state of the science on the CBR approach and tabulated estimated and measured CBRs for several mode-of-action categories (Table 1). As they noted, the tabulated data were not intended to be a comprehensive compilation of all available data but rather a summary of ranges of values considered to be representative of CBRs for chemicals in various mode-of-action classes. The data summary and evaluation by McCarty and Mackay (1993) suggested that nonpolar narcotics generally had the highest CBRs and showed limited variability, that polar narcotics had lower CBRs than nonpolar narcotics, and that chemicals with more specific modes of action generally had the lowest CBRs. van Wezel and Opperhuizen (1995) examined the

Table 1. Ranges of CBRs for the acute and chronic toxicity of chemical classes.

Chemical class/ Mode of action	Lethal effects (mmol/kg)	Sublethal effects (mmol/kg)
Narcotics	1.7 to 8 (nonpolar)	0.2 to 0.8 (nonpolar)
	0.6 to 1.9 (polar)	0.2 to 0.7 (polar)
Excitatory agents	0.1 to 0.3[a]	1.5×10^{-4} to 9×10^{-2}
AChE inhibitors	5×10^{-2} to 2.7	3×10^{-3} (data for one chemical)
Reactives/irritants	9.4×10^{-3} to 13	No data
CNS seizure agents	4.8×10^{-5} to 1.7×10^{-2}	5×10^{-4} to 1.5×10^{-2}
Ah-receptor agonists	1.5×10^{-7} to 4×10^{-5}	3×10^{-7} to 8×10^{-6}

CBR, critical body residue.
[a]Range excludes one possible outlier value for 2,4-dinitrophenol of 1.5×10^{-3} mmol/kg.
Source: Summarized by McCarty and Mackay (1993).

experimental and theoretical support of the CBR approach summarized by McCarty and coworkers and concluded that lethal residues of narcotics were relatively constant and varied from 2 to 8 mmol/kg. McCarty and Mackay (1993) and McKim and Schmieder (1991) speculated that there are relatively narrow ranges of CBRs that are characteristic of a particular mode of action. However, these scientists also cautioned that the CBR approach may not apply to all modes of action, such as those with receptor-mediated and surface-acting toxicity.

B. Uncertainties in the CBR Approach

Despite its potential advantages, the CBR approach has not been comprehensively evaluated across a broad range of chemicals and multiple mode-of-action classes. Barron et al. (1997) previously noted that most of the theoretical and experimental support for the CBR approach has been limited to a relatively small number of narcotic chemicals, and most experiments have been relatively short-term exposures conducted with small fish. Barron et al. (1997) reviewed structure–activity relationships for CBRs in relation to mechanism of action, toxicity endpoints, mixture toxicity, exposure dependence, dose dependence, chemical transformation, and chemical activation, and concluded that additional evaluation was needed before CBRs could be broadly applied in ecological assessments. Sources of uncertainty in the CBR approach include extrapolating data between species, exposure regimes, and chemicals (Barron et al. 1997). Additional uncertainty and potential variability in CBRs involve the correct assignment of chemical mode of action, and effects of metabolism, excretion, and tissue disposition of chemicals in aquatic organisms.

III. Methodology for Evaluating CBRs

This evaluation of contaminant tissue residues and adverse effects in aquatic organisms is based on an examination of existing data. The information reviewed included both primary scientific literature and a recently developed comprehensive database summarizing exposure concentrations, tissue residue concentrations, and general toxicity endpoints (Jarvinen and Ankley 1999). The Jarvinen and Ankley (1999) database contains data from more than 500 research articles and includes effects from approximately 200 organic and inorganic chemicals. The majority of the studies contained in the database were not designed explicitly to measure CBRs, and thus a range of potential relationships between tissue residue and adverse effect may exist (see Fig. 1). Also, the endpoints presented by Jarvinen and Ankley (1999) were variable in the level of effect, such as the percentage mortality observed at a measured tissue concentration; thus, a given tissue residue is not numerically associated with a specific effect level. Despite these limitations, the Jarvinen and Ankley (1999) database provided a useful and reasonable approximation of the range of tissue residues for a range of adverse effects for many compounds.

The original information source was inspected when determining the minimum and maximum values in the CBR range determined from the Jarvinen and Ankley (1999) database and supplemented with newer data from the primary literature. Values from multiple species, exposure regimes, and analytical methods were included in the range of CBRs. A large number of individual studies were also examined to determine the consistency of CBRs across species of aquatic organisms, experimental methods, and exposure regimes. Consistent with the CBR approach (McCarty 1987), reported tissue concentration data (i.e., mg/kg) were converted to molar concentrations (mmol/kg) using the molecular weight of the compound. This conversion is considered appropriate for narcotic chemicals (McCarty and Mackay 1993), but its applicability to other modes of action is unknown.

The consistency of CBRs for selected individual chemicals was evaluated by first compiling data on tissue concentrations of chemicals causing mortality to determine if CBRs were consistent between species, exposure temperatures, and other environmental, chemical, and biological variables. When studies produced consistent CBRs using different environmental, chemical, or biological variables, we concluded that the particular variable did not influence the CBR. Substantial differences in CBRs between studies provided evidence that particular environmental, chemical, or biological variables did influence the relationship between bioaccumulation and toxicity and that the chemical did not consistently accumulate to a critical level to produce toxicity.

The consistency of CBRs within chemicals having the same mode of action was evaluated by segregating tissue residue data into one of the eight mode-of-action classes for organic and inorganic chemicals identified in Table 2. The classification schemes for organic chemicals followed that of McCarty and Mackay (1993) and Russom et al. (1997). If the mode of action of the chemical was uncertain, the value was excluded from further analysis. Single extreme low or extreme high values (two orders of magnitude below or above the next value) were excluded from the reported range of CBRs. We also evaluated the range of CBRs for major chemical subgroups within a mode-of-action category (Russom et al. 1997). The chemical subgrouping was based on known mode-of-action subcategories (e.g., nonpolar versus polar narcotics) or major differences in chemical structure, such as organochlorine and pyrethroid CNS seizure agents.

IV. Narcotic Chemicals
A. Background

Narcotics are a structurally diverse group of low molecular weight chemicals that cause hypoactivity and anesthesia (i.e., depression of locomotor and sensory functions). They are the largest class of synthetic organic chemicals and include alkanes, benzenes, simple alcohols, phenols, ketones, and esters. Narcotic chemicals are used as solvents and chemical intermediates and are components of paints and adhesives. They are not reactive, and generally have moderate water

Table 2. Chemical classes and modes of action evaluated.[a]

Chemical class	Mode of action	Structural features	Example chemicals
Narcotics	Narcosis (anesthesia) Nonspecific action	Small molecules, diverse structure, nonreactive	Phenols, ketone, esters, alcohols, benzenes
Excitatory agents	Oxidative phosphorylation uncoupling	Poly halo- and nitrobenzene derivatives	Pentachlorophenol, dinitrophenols
AChE inhibitors	AChE inhibition	Organophosphorus and carbamate pesticides	Chlorpyrifos, terbufos, carbaryl
Reactives/irritants	Irritation/damage to membranes and nerve tissue	Electrophilic structures (e.g., reactive double bonds)	Aldehydes, alkenes, alkynes
CNS seizure agents	Nervous system interaction	Organochlorine and pyrethroid insecticides	DDT, fenvalerate
Ah-receptor agonists	Ah-receptor binding	Planar chlorinated aromatic compounds	TCDD, planar PCBs
Heavy metals	Ionoregulation, respiration, cell damage	Free cation and complexes with simple anions	Copper, cadmium, zinc, arsenic
Organometals	Nerve tissue damage	Metal–organic complexes	Methylmercury, tributyltin

[a] Chemical classes and mode-of-action categories were adapted from McCarty and Mackay (1993) and Russom et al. (1997), and were expanded to include heavy metals and organometals.

solubility, high soil mobility, high volatility, and low bioaccumulation. They partition primarily in the vapor phase and rapidly equilibrate between exposure water and tissue.

Narcotic chemicals have rapidly reversible anesthetic effects and do not appear to cause cumulative injury. Their mode of action is a nonspecific and reversible interaction with cellular lipids and proteins (van Wezel and Opperhuizen 1995). Potential mechanisms of narcosis include (a) changes in membrane fluidity, (b) altered membrane protein function through interaction with membrane lipids or direct protein binding, and (c) disruption of nerve function (van Wezel and Opperhuizen 1995; van Wezel et al. 1996). Narcotic chemicals have been grouped into two mode-of-action subcategories (i.e., nonpolar narcotics and polar narcotics) on the basis of chemical structure and general degree of toxicity (Russom et al. 1997). Russom et al. (1997) included a third subcategory of narcosis for ester narcotics, such as benzoates and acetates. However, ester narcotics are not treated separately from nonpolar narcotics in the CBR approach, and data for these chemicals are limited. In this review, only nonpolar and polar narcotics are considered. Nonpolar narcotics include the majority of chemicals causing narcosis and are characterized as having the lowest toxicity and the highest CBRs (Vaes et al. 1998). Polar narcotics are more polar chemicals that elicit initial excitatory responses followed by narcotic-like depression (Bradbury et al. 1989). Polar narcotics include phenol and phenolic compounds substituted with one to two electron-withdrawing groups such as nitro or halo substituents. Phenolic compounds containing additional electron-withdrawing groups may act as excitatory agents rather than as narcotics. The polar narcotics evaluated in this review included phenol, mono- and dihalogenated phenols, and nitrophenol. The excitatory agents evaluated in this review (Section V.A) included trichlorophenol, tetrachlorophenol, pentachlorophenol, and dinitrophenol.

B. Variability of CBRs

There is both theoretical and experimental support for the presence and consistency of CBRs for narcotics, but research has been limited to a relatively small subset of chemicals such as chlorophenols, chlorobenzenes and anilines, and chlorinated alkanes (van Wezel and Opperhuizen 1995). Comprehensive evaluation of the Jarvinen and Ankley (1999) database and primary literature for a broad range of narcotic chemicals shows that there is considerable variation in CBRs between studies of the same chemicals or chemicals with nearly identical structures. For example, the lethal body residues of the nonpolar narcotic 1,2-dichlorobenzene varied by one order of magnitude between different studies: 2–3 mmol/kg (van Hoogen and Opperhuizen 1988), 0.42 mmol/kg (Pawlisz and Peters 1993a), and 1–4 mmol/kg (van Wezel et al. 1996). Lethal body residues of 1,4-dichlorobenzene differed more than 20 fold, ranging from 0.39 to 8 mmol/kg (van Wezel et al. 1996). The lethal body residue of the fluorinated isomer 1,4-difluorobenzene was 28 mmol/kg (Sijm et al. 1993). Hence, a 72-

fold difference in CBRs has been measured for two dihalogenated benzenes derived from just four studies with *Daphnia magna*, guppies (*Poecilla reticulata*), and fathead minnows. Considerable variation in reported lethal body residues was also observed for the 1,2,3- and 1,2,4-isomers of the nonpolar narcotic trichlorobenzene. CBRs in these same test species varied by one order of magnitude between several studies: 2–3 mmol/kg (van Hoogen and Opperhuizen 1988), 21 mmol/kg (Pawlisz and Peters 1993a), 5.3–20 mmol/kg (de Maagd et al. 1997), and 3–6 mmol/kg (van Wezel and Jonker 1998). This variation suggests that relatively high variation in CBRs can be observed even between studies of the same chemical. With the data available, the cause of the variation (i.e., chemical, biological, or environmental factors) could not be determined between studies of a single chemical or between studies of similar chemicals.

Examination of the Jarvinen and Ankley (1999) database shows that tissue residues of narcotics (both polar and nonpolar) that reduce survival range from 0.009 mmol/kg for 3-chlorocresol to 450 mmol/kg for butanol (tested in brown trout, *Salmo trutta*, and *Daphnia magna*, respectively). This 50,000-fold range in tissue residues incorporates 144 reported database values, which were mostly for benzenes, phenol, chlorophenols, nitrophenols, alkyl naphthalenes, ketones (e.g., acetone), and tetrachlorethanes. More specifically, the tissue residue data for nonpolar narcotics incorporated 80 reported values and ranged from 0.032 to 450 mmol/kg (Jarvinen and Ankley 1999). Tissue residues of polar narcotics that reduce survival ranged from 0.009 to 4.9 mmol/kg and comprised 64 reported values (Jarvinen and Ankley 1999). Tissue residues associated with adverse effects were generally lower for polar narcotics than for nonpolar narcotics, but the data ranges overlapped substantially. Large variation was also evident between narcotic chemicals within the same study using consistent experimental conditions. For example, Fig. 2 shows that lethal body residues exhibited a 3,500-fold range in *Daphnia magna* exposed to each of 10 narcotics, from 0.1 mmol/kg for 2-methylnaphthalene to 350 mmol/kg for butanol (Pawlisz and Peters 1993a,b). The variability in tissue residues associated with adverse effects, both within and between chemicals, indicates that the CBR approach may have limited application even for narcotic chemicals unless factors such as species sensitivity, biological variability, and exposure conditions are considered.

C. Determinants of CBRs

According to CBR theory, lethal tissue residues of narcotic chemicals should be approximately 2 mmol/kg tissue and should be relatively independent of chemical, biological, and environmental factors. However, critical evaluation of the available data shows that the CBRs of narcotic chemicals can show substantial variability. Determinants of CBRs include chemical structure, biological factors such as biotransformation and lipid content, and environmental factors such as ultraviolet radiation (UV) and pH.

The CBRs of phenolic compounds summarized by McCarty et al. (1993)

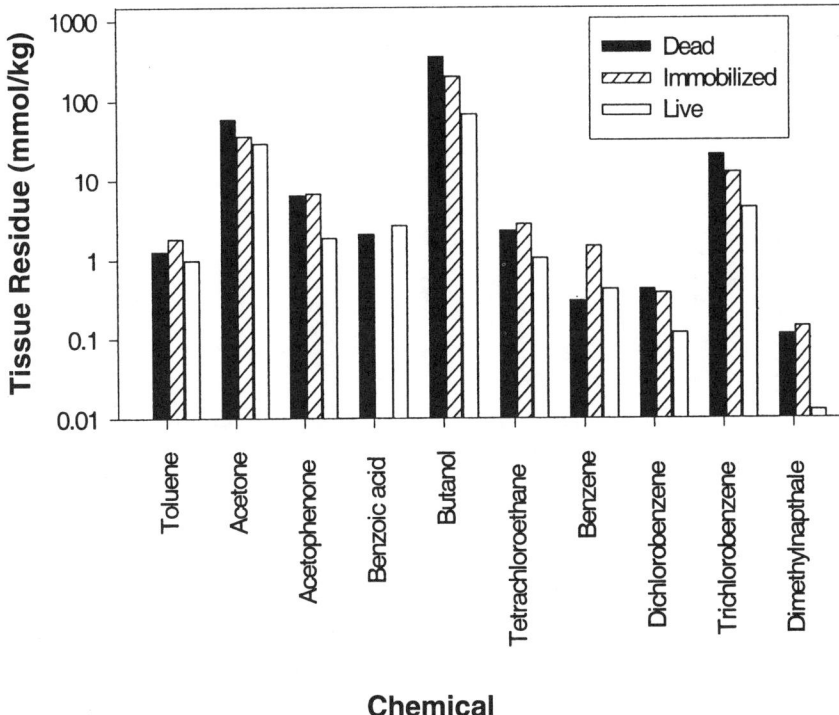

Fig. 2. Tissue residues of 10 narcotic chemicals in dead, immobilized, and live *Daphnia magna*. (Data from Pawlisz and Peters 1993a,b.)

exhibited approximately three orders of magnitude range in CBRs, similar to the variability in $LC_{50}s$. The addition of halogen groups appears to lower CBRs for phenols (Fig. 3), and consideration of substituent groups may provide a means of reducing variability in CBRs across individual compounds (McCarty et al. 1993; Di Toro et al. 2000). A reduction in CBR appears to be related to a change in mode of action from narcosis to excitation. Mono- and dichlorophenols are considered to be polar narcotics, whereas trichlorophenol, tetrachlorophenol, and pentachlorophenol are considered to be excitatory agents (McCarty et al. 1993; Penttinen and Kukkoken 1998). Phenols with electron-donating (e.g., alkyl) groups have higher CBRs and are considered nonpolar narcotics (McCarty et al. 1993). Some nonpolar narcotics such as halogenated benzenes have been shown to produce CBRs that appear to be relatively independent of molecular shape, molecular size, and octanol–water partition coefficient (K_{ow}) (Sijm et al. 1993). However, most studies (e.g., Connell and Markwell 1992) have not evaluated broad groups of structurally diverse narcotics, and the variation in CBRs was not systematically evaluated to determine whether there was a relationship between chemical structure and tissue residues.

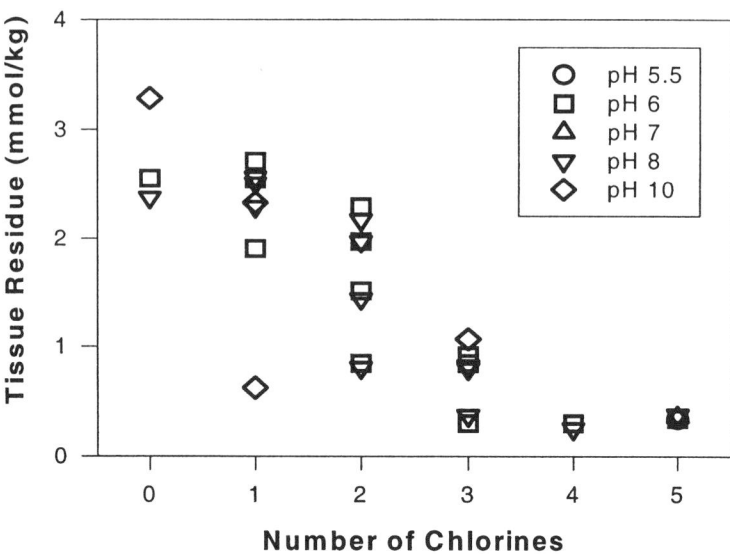

Fig. 3. Tissue residues at death in goldfish (*Carassius auratus*) exposed to phenol, monochlorophenols, dichlorophenols, trichlorophenols, 2,3,4,6-tetrachlorophenol, and pentachlorophenol. Exposures were conducted for 5 hr at pH 5.5–10. Each symbol represents the mean tissue residue of several fish samples. (Data from Kishino and Kobayashi 1995.)

Biotransformation by aquatic organisms may account for some of the variability in narcotic CBRs. For example, biotransformation of three PAHs in *Hyalella azteca* resulted in greater accumulation of total residues (parent + metabolites) than would be expected if no biotransformation had occurred (Lee et al. 1999). Tissue residues of fluoranthene in the amphipod *Leptocheirus plumulosus* were made up of approximately 30% metabolites (Kane Driscoll et al. 1998). Tissue residues of chemicals that are biotransformed may appear less toxic because of a higher apparent CBR if tissue concentration measurements are based on total residues. This discrepancy occurs because both parent chemical and less toxic metabolites are quantified rather than only the parent compound (McCarty et al. 1993). Biotransformation may be increased by prior chemical exposure, and the induction of metabolic activity may be dependent on the species and exposure conditions.

Different species or individuals of a species may have different concentrations of lipids, which can contribute to CBR variability. CBRs for narcotics should increase in direct proportion to the lipid content of the organism (McCarty et al. 1992) because the mechanism of toxicity for narcotics appears to involve nonspecific binding to lipoidal tissues. Additionally, higher lipid content may act to lower the concentration of chemical in the target tissue. A correlation between increasing lipid content and higher CBRs has been shown by several researchers (van Wezel et al. 1995; de Maagd et al. 1997), and has become the basis of the target lipid model of DiToro et al. (2000) and DiToro and McGrath (2000).

In a recent critical review of CBRs of 33 species exposed to 156 narcotic chemicals, DiToro et al. (2000) also found CBRs to be highly variable. These authors concluded that nonpolar narcotic chemicals had variable chemical potency, with aliphatics, ethers, alcohols, and aromatics having similar potency, and ketones, halogenated chemicals, and PAHs having much higher potency. DiToro et al. concluded that variability in CBRs for narcotics could be reduced by recognizing this differential potency, by incorporating the percent lipid concentration of organisms into a QSAR model, and by explicitly recognizing differential species sensitivity. Although the resulting target lipid model of DiToro et al. accounted for much of the variation of CBRs for the narcotic chemicals examined, there is still considerable variation associated with observed body residues. On average, the target lipid model was found to increase the accuracy of predicted mean CBRs for individual species, but individual measurements of nonpolar narcotic chemicals within a species still varied by a factor of 10 or more (DiToro et al. 2000).

Environmental exposure conditions such as differences in temperature, pH, and salinity may alter CBRs through their influence on metabolism, chemical bioavailability, and degree of chemical ionization. For example, higher temperatures were shown to reduce the CBR for 1,2,4-trichlorobenzene (van Wezel and Jonker 1998), and the CBR of 4-nitrophenol was influenced by both the temperature and pH of the solution (Howe et al. 1994). Alternatively, pH had only a minor effect on the lethal body residues of 12 chlorophenols (Kishino and Kobayashi 1995). Although data are limited, the route of exposure may be a less important source of variability of CBR variability. For example, the CBR in amphipods exposed to fluoranthene in sediment were 0.3 mmol/kg, similar to the 0.4 mmol/kg in amphipods exposed in a water-only medium (Kane Driscoll et al. 1998).

Photoenhanced toxicity, which is a 2- to 1000-fold increase in toxicity observed under UV compared to fluorescent lighting (minimal UV), can alter the CBRs for specific PAH compounds with three to five benzene or substituted benzene rings. For example, photoactivated fluoranthene at 0.12 mmol/kg caused 98% mortality, whereas nonphotoactivated fluoranthene at 1.25 mmol/kg had no effects on survival (Ankley et al. 1995). Most CBR data have been derived from laboratory studies with minimal UV, and the CBR ranges summarized in this review do not include chemicals that exhibit photoenhanced toxicity. CBRs of photoactivated PAHs are dependent on the degree, duration, and spectrum of UV exposure, and not only on the tissue residue of the PAH (Barron et al. 2000). Application of the CBR approach to photoactivated chemicals may be problematic because the UV exposure of the organisms must be determined in addition to tissue residues.

Empirically derived CBRs have been found to vary considerably for many narcotic chemicals. Causes of this variability include differences in chemical structure, species responses, and exposure conditions. The variability in the reported CBRs may also relate to the fact that relatively few of the available data were derived from studies aimed specifically at determining CBRs. Also, there is some uncertainty associated with the mode of action of the chemicals considered to be narcotics.

V. Nonnarcotic Organic Chemicals

Chemicals with more specific modes of action include excitatory agents, AChE inhibitors, reactive chemicals, CNS seizure agents, and Ah receptor agonists. With the exception of a few investigations of excitatory agents, research specifically evaluating the chemical, biological, and environmental determinants of CBRs for nonnarcotic chemicals is limited. The CBR concept was developed for narcotic chemicals, and was only postulated to apply to other chemical groups.

A. Excitatory Agents

Excitatory agents are a smaller group of substituted phenolic chemicals that cause hyperactivity and overreaction to outside stimuli. Generally, this group is composed of chemicals with a single aromatic ring (e.g., phenols, anilines, pyridines) containing multiple electron-withdrawing substituents (e.g., halogens and nitro groups). These chemicals include some chlorophenols such as pentachlorophenol (PCP) and some nitrophenols such as dinitrophenol that are used as solvents, chemical intermediates, and biocides. Trichlorophenols were grouped in the excitatory agent category in agreement with the evaluations of Penttinen and Kukkoken (1998), although these chemicals may also act as polar narcotics (Russom et al. 1997). Chlorophenols, particularly PCP, have also been widely used as pesticides and wood preservatives. The excitatory agents generally have lower water solubility, higher affinity for sediments and soil, and less volatility, and are more persistent in the environment than narcotic agents. The mechanism of action of the excitatory agents is an uncoupling of oxidative phosphorylation (i.e., dissociation of the electron transport process from the generation of ATP).

In general, CBRs for excitatory agents appear to be lower than those for polar narcotics, despite their structural similarities. The number and type of electron-withdrawing groups on a phenol determine its mode of action and thus influence the magnitude of the CBR. On average, phenols with four and five chlorines are considered excitatory agents and have been found to cause death at tissue residues of approximately 0.3 mmol/kg, whereas monochlorophenols and dichlorophenols are considered narcotics and generally caused death at tissue residue concentrations greater than 1 mmol/kg (see Fig. 2). The range of tissue residues associated with mortality is highly variable between several excitatory chemicals. Tissue residues of PCP associated with reduced survival in fish and invertebrates range from 0.0004 mmol/kg (Fisher et al. 1999) to 0.9 mmol/kg (Landrum and Dupuis 1990), which represents a 2300-fold range in CBRs for just one chemical. However, within related taxa such as fish, CBRs for PCP were similar and ranged from 0.05 mmol/kg for rainbow trout (McKim and Schmieder 1991) to 0.3 mmol/kg for fathead minnows (Arthur 1991; Hickie et al. 1995) to 0.4 mmol/kg for goldfish (*Carassius auratus;* Kishino and Kobayashi 1995). For 2,4-dinitrophenol, CBRs from the same study ranged from 0.11 to 4.4 mmol/kg, which is a 40-fold range in the amphipod *Gammarus pseudolimnaeus*, and from 0.034 to 0.36 mmol/kg, a 10-fold range, in rainbow trout depending on exposure pH and temperature (Jarvinen and Ankley 1999).

Exposure temperature and pH both appear to account for some of the variability of CBRs for excitatory agents, whereas other environmental variables appear to have less influence. In a study of amphipods exposed to 2,4-dinitrophenol, CBRs at pH 6.5 were 30 times lower than those at pH 8.5, whereas at a constant pH, CBRs were within a sixfold range at different temperatures (Jarvinen and Ankley 1999). A combination of higher pH and lower temperature substantially increased PCP CBRs in zebra mussels, *Dreissena polymorpha* (Fisher et al. 1999) (Fig. 4). In the Fisher et al. (1999) study, CBRs for 50% mortality increased eight times with increasing pH and decreasing temperature, whereas $LC_{50}s$ (µmol/L) increased 380 times. In another study of chlorophenols, pH had only a minor effect on the CBRs for mortality in goldfish (Kishino and Kobayashi 1995). CBRs were shown to be independent of exposure time for PCP and for 2,4,5- and 2,4,6-trichlorophenol (Klee 1998). Similarly, few differences in CBRs for PCP were observed between constant exposure concentrations and pulsed exposure concentrations (Hickie et al. 1995). However, because of the limited research on excitatory agents, additional data are required to ex-

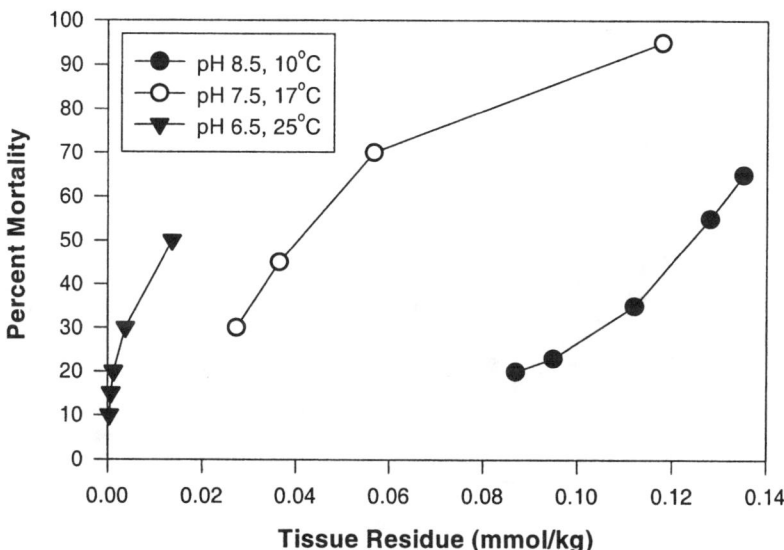

Fig. 4. Relationship between tissue residues of PCP and mortality of zebra mussels (*Dreissena polymorpha*) when exposed at different pH values and temperatures. (Data from Fisher et al. 1999.)

plore the effects of these and other environmental, chemical, and biological variables on CBRs of these chemicals.

Limited data were available for assessing the sublethal effects of CBRs for excitatory agents. In general, sublethal effects were observed at lower body residues than lethal effects, but with similar variability (Jarvinen and Ankley 1999). Figure 5 shows that fathead minnows exposed to PCP for 28 days exhibited a general trend of decreasing growth with increasing tissue residues, but data for individual fish were highly variable (Arthur 1991). Penttinen and Kukkonen (1998) quantified the disturbance of normal energy metabolism as the rate of heat dissipation by the excitatory agents 2,4-dichlorophenol (DNP), 2,4,5-trichlorophenol (TCP), and PCP in two species of freshwater invertebrates: chironomid larvae (*Chironomus riparius*) and an oligochaete worm (*Lumbriculus variegatus*). Heat output was unaffected at low body residues, but increased linearly with tissue residue concentrations when tissue residues were in the range of 0.2 to 2 mmol/kg (Penttinen and Kukkonen 1998).

Overall, data on the application of CBR approaches to excitatory agents suggest that CBRs may be relatively constant for some compounds such as PCP, although differences in species sensitivity are apparent. However, CBRs appear to vary as a function of environmental exposure factors such as temperature and pH, which limit the environmental applications of the approach. Additional focused research on CBRs for key compounds is necessary to more adequately evaluate application of the CBR concept for excitatory agents.

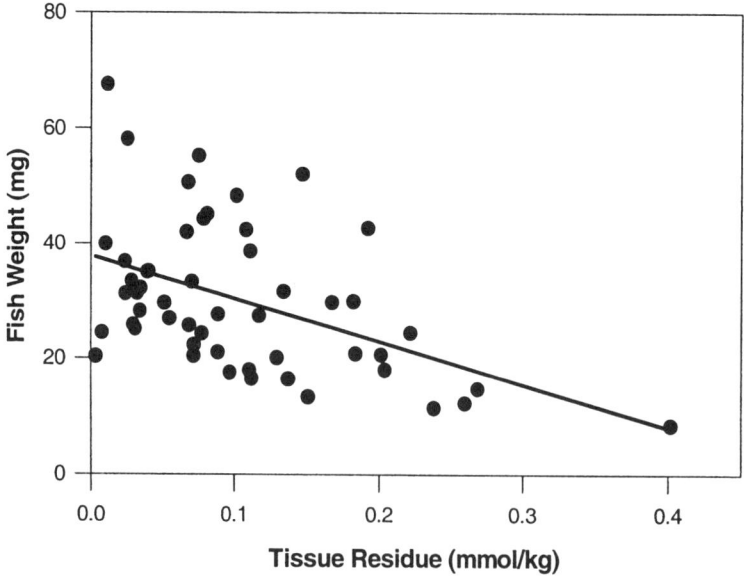

Fig. 5. Tissue residues of pentachlorophenol (PCP) and growth of fathead minnows (*Pimephales promelas*) after 28 d exposure. (Data from Arthur 1991.)

B. Acetylcholinesterase Inhibitors

Acetylcholinesterase (AChE) inhibitors are organophosphate insecticides (OPs) such as terbufos, trichlorfon, and chlorpyrifos, and carbamate insecticides such as carbofuran and carbaryl. AChE inhibitors cause toxicity through inhibition of AChE, an enzyme that degrades the neurotransmitter acetylcholine. In contrast to the nonspecific toxicity of narcotics, AChE inhibitors cause toxicity through a specific receptor-mediated mechanism (Legierse et al. 1999). OPs and carbamates are applied in agricultural and residential uses, and can also be accidentally introduced into the environment through overspray, aerial drift, or runoff. They are typically produced as esters, amides, or thiol derivatives, which affect their physical and chemical properties and environmental fate. In general, they have low solubility in water, high soil binding, and low volatility. OPs and carbamates are less persistent and less bioaccumulative than the chlorinated insecticides such as DDT and dieldrin. Carbamate insecticides are generally less persistent in the environment and less hydrophobic than OPs.

Tissue residues of AChE inhibitors associated with reduced survival appear to be highly variable, ranging from 0.00004 mmol/kg for trichlorfon to 29 mmol/kg for dichlofenthion (Deneer et al. 1999; Jarvinen and Ankley 1999). This range encompasses 115 reported values with a 725,000-fold range in CBRs, and includes 17 OPs and 1 carbamate insecticide. Tests were conducted with multiple species, including amphipods (*Gammarus psudolimnaeus*), grass shrimp (*Palaemonetes pugio*), sheepshead minnows (*Cyprinodon variegatus*), rainbow trout, and guppies. However, the majority of the residue data were for the OPs terbufos and trichlorfon. One extreme low value (10^{-6} mmol terbufos/kg) was excluded from the range of CBRs because it was 28 fold lower than any other value found in the database (Jarvinen and Ankley 1999).

High variation in lethal body residues of AChE inhibitors was observed both within single studies of the same chemical and across studies of multiple chemicals. Lethal residues of the OP chlorthion in pond snails (*Lymnaea stagnalis*) ranged from 0.015 to 0.63 mmol/kg, which is a 40-fold range of CBRs. There was no apparent correlation between lethal tissue residues and lipid content (Legierse et al. 1999). Similarly, lethal residues of 13 OPs in guppies exhibited a 2600-fold range in CBRs of 0.011 to 29 mmol/kg, with no apparent correlation with K_{ow} or exposure duration (Deneer et al. 1999).

Variability in lethal body resides of AChE inhibitors may be attributable to species differences and exposure duration, although further research may indicate influences of other variables. Differences in intrinsic species sensitivity to AChE inhibitors is caused by different kinetics of AChE binding such as receptor number or affinity or different rates of biotransformation (e.g., detoxification versus bioactivation rates; Barron and Woodburn 1995). Other research has shown that lethal tissue residues are lower at lower exposure concentrations and longer exposure durations (Legierse et al. 1999). Using this relationship, Legierse et al. (1999) developed a time-dependent model of the toxicity of OPs that incorporates time-dependent metabolic bioactivation and receptor binding and AChE inhibition. Although some AChE inhibitors require metabolic activation to elicit toxicity, studies have not clearly shown that metabolic activation affects CBR values. For example, similar CBRs were reported in grass shrimp and sheepshead minnows for terbufos, which requires metabolic activation, and trichlorfon, which does not (Jarvinen and Ankley 1999).

The high variation of tissue residue concentrations associated with lethality to AChE inhibitors leads to the conclusion that the CBR concept does not apply to this class of chemicals. This conclusion is strengthened by the observations of differences in species sensitivity and differences in lethal tissue residues attributable to exposure duration. The nearly 1,000,000-fold range in lethal tissue residues for AChE inhibitors makes environmental applications of the CBR approach problematic for this class of chemicals.

C. Reactive Chemicals

Reactive chemicals are a small group of low molecular weight compounds such as aldehydes, unsaturated aliphatics (e.g., alkenes, alkynes), and alcohols (e.g., propargylics). These chemicals contain structural features (e.g., reactive double bonds) that allow electrophilic or proelectrophilic reaction with nucleophilic structures such as amino and thiol groups on cellular macromolecules. They can react with DNA, enzymes, and other proteins, causing irritation or damage to mucous membranes and nerve tissues. The reaction between the chemical and tissue alters the functionality of cellular components, resulting in a diversity of behavioral and neurotoxic responses, including narcosis, hyperactivity, and convulsions (Russom et al. 1997). Sources of reactive chemicals include biocides such as acrolein; chemical intermediates from chemical production; products of organic pyrolysis such as automobile emissions, manufacturing, pulp and paper effluents; solvents; and industrial fluids. Because of their various chemical structures, reactive chemicals can have low (e.g., chlorinated alkenes) to high (e.g., acrolein) water solubility, moderate volatility, and low to high soil binding.

Tissue residues of reactive chemicals associated with reduced survival range from 0.094 mmol/kg for acrolein to 13 mmol/kg for benzaldehyde (McKim and Schmieder 1991; Jarvinen and Ankley 1999). This range encompasses only eight reported values with a 140-fold range in CBRs, with data for only acrolein, benzaldehyde, and hexachloro-1,3-butadiene. Other chemicals including PAHs may be bioactivated to reactive metabolites such as dihydrodiol epoxides of benzo(a)pyrene but were not considered as reactive chemicals in this review.

Although only very limited data on CBRs for reactive chemicals are available, the CBR approach does not appear applicable to this class of chemicals. Verhaar et al. (1999) presented theoretical arguments that CBRs will not be constant for chemicals with a mechanism of action that includes irreversible or partially irreversible effects. Verhaar et al. also argued that CBRs will not be constant for reactive and receptor-mediated chemicals, and that the CBRs may decline with exposure time. McKim and Schmieder (1991) concluded that acrolein and benzaldehyde were irritants that acted at the surface of the gills and that whole-body residues were not important in determining their toxicity. An additional complication is that the acute effects of more hydrophobic reactive chemicals may also be associated with narcosis, possibly because of high affinity for membrane lipids (Hermens 1990). This affinity may explain some of the relatively high lethal tissue residues for some reactive chemicals (McKim and Schmieder 1991; Jarvinen and Ankley 1999).

D. CNS Seizure Agents

The CNS seizure agents include two general classes that cause tremors and convulsions by damaging nerve tissues: the organochlorine pesticides (OCs) and the pyrethroid insecticides such as fenvalerate, permethrin, and cypermethrin. The OCs include ethanes such as DDT, cyclodienes such as chlordane and dieldrin, hexachlorocyclohexane (lindane), the chlorinated camphene toxaphene, mirex, and chlordecone (kepone). The OCs are a group of chlorinated broad-spectrum insecticides produced commercially beginning in the 1940s for agricultural and residential use. Most uses of OCs have been banned worldwide because of extreme environmental persistence (half-lives of months to years) and extensive biomagnification in the food web. Characteristics of OCs include high lipid solubility, low water solubility, and high affinity for sediments and soils. They are mobilized by soil runoff, airborne dusts, and suspended sediments. Pyrethroids are substantially less persistent in the environment than the OCs and are less hydrophobic.

Data in Jarvinen and Ankley (1999) indicate that tissue residues associated with mortality were highly variable for CNS seizure agents. CBRs for reduced survival ranged from 0.00002 mmol/kg for cypermethrin to 1.1 mmol/kg for DDT. This range incorporated 125 reported values and demonstrates a 55,000-fold range in CBRs. Tests were performed with multiple species, including pink shrimp (*Penaeus duorarum*), white shrimp (*Penaeus setiferus*), midges (*Chironomus riparius*), several crab species, sheepshead minnows, *Daphnia magna*, chinook salmon (*Oncorhynchus tshawytscha*), brook trout (*Salvelinus fontinalis*), goldfish, fathead minnows, and other fish species. High variability was observed even within structurally similar chemicals. Lethal tissue residues for OCs ranged from 0.00007 to 1.1 mmol/kg, and those for pyrethroids ranged from 0.00002 to 0.012 mmol/kg (Jarvinen and Ankley 1999).

Factors contributing to the broad range in lethal tissue residues appear to include chemical structure, species differences, exposure dynamics, and tissue-specific accumulation. Different chemicals and stereoisomers of the same chemical have been shown to have substantially different CBRs. For example, Bradbury et al. (1987) observed more than a 30-fold difference in the lethal body residues of isomers of fenvalerate in fathead minnows. The $2S$ stereoisomer of fenvalerate caused mortality within 32 hr of exposure and tissue concentrations ranged from 0.0005 to 0.004 mmol/kg, whereas the $2R$ stereoisomer did not cause mortality within 48 hr and average tissue residues were 0.017 mmol/kg.

Figure 6 shows that across several studies the lethal tissue residues for DDT in 11 species of aquatic invertebrates and fish varied by a factor of 10,000 (Jarvinen and Ankley 1999). Exposure dynamics such as continuous versus pulsed exposures have also been shown to affect both lethal and sublethal CBRs of CNS seizure agents (Fig. 7). Tissue residues associated with reduced survival can also vary substantially in individual tissues. For example, Fig. 8 shows that no-effect concentrations of DDT in intestine and liver were higher than lethal levels of DDT residues in brain, heart, spleen gill, muscle, and carcass. These data indicate that application of the CBR concept to the overall class of CNS seizure agents is problematic.

E. Aryl Hydrocarbon Receptor Agonists

Aryl hydrocarbon (Ah) receptor agonists are compounds that have high affinity for the Ah receptor, which is a protein complex in the cell nucleus. Chemical binding to the Ah receptor elicits of a complex cascade of biochemical and physiological responses including induction of the P-450 oxidative enzyme system. Ah-receptor agonists include halogenated aromatics such as PCBs, dibenzo-p-dioxins (PCDDs), and dibenzofurans (PCDFs). The potency of these polycyclic aromatic compounds is determined by the chemical structure such as position of chlorine atoms and molecular conformation including the ability to form a planar configuration. For example, the most potent PCBs, PCDDs, and PCDFs have outward (meta, para) halogen substitutions rather than inward (ortho) substitutions, which increases binding affinity for the Ah receptor. Al-

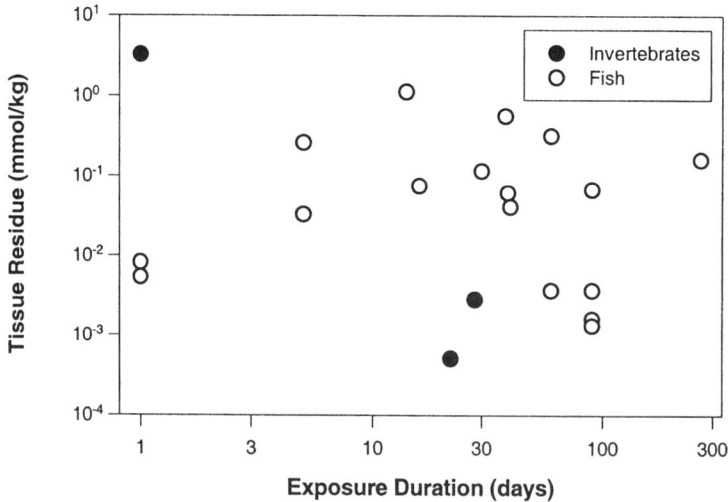

Fig. 6. Tissue residues for reduced survival in 11 species of aquatic invertebrates and fish exposed to DDT. Exposure duration ranged from 1 to 266 d. (Data from Jarvinen and Ankley 1999.)

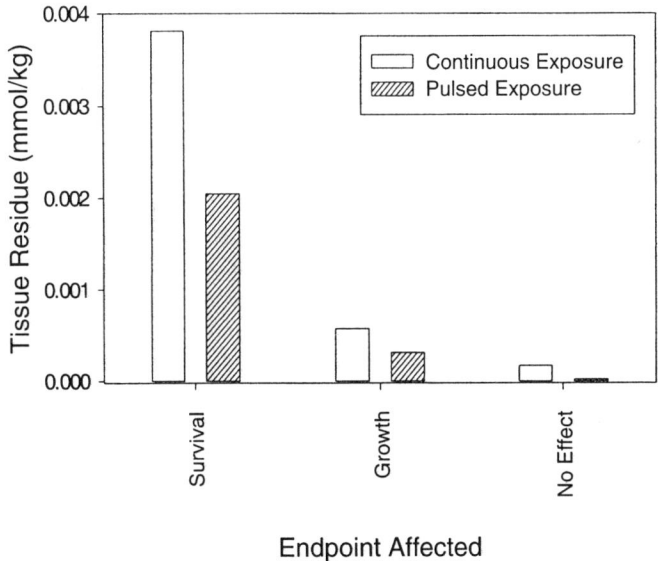

Fig. 7. Tissue residues of fenvalerate in rainbow trout (*Oncorhynchus mykiss*) exhibiting reduced survival, growth, or no adverse effects during continuous or pulse exposures. (Data from Jarvinen and Ankley 1999.)

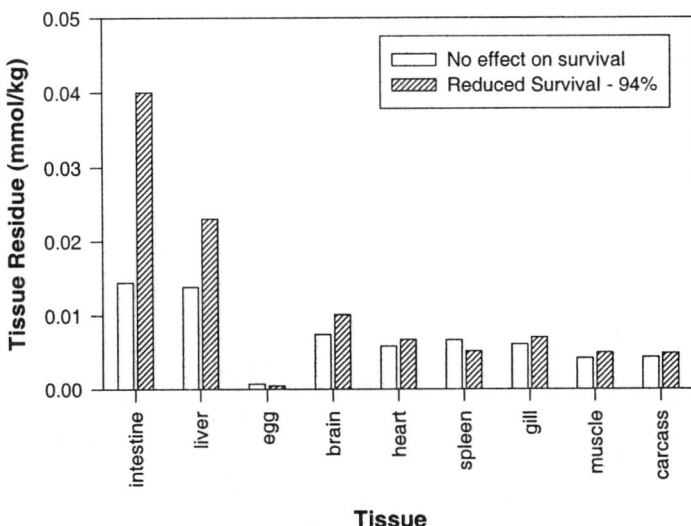

Fig. 8. Tissue residues of mummichogs (*Fundulus heteroclitus*) showing no effects and reduced survival following exposure to waterborne DDT for 1–2 d. (Data from Jarvinen and Ankley 1999.)

though these chemicals have specific mechanisms of action, some Ah-receptor agonists such as PCBs may exhibit a narcosis mode of action during short-term exposures (van Wezel and Opperhuizen 1995).

PCDDs and PCDFs, including TCDD (2,3,7,8-tetrachlorodibenzo-p-dioxin), are contaminants produced in chemical (e.g., Kraft process pulp and paper mills, chemical industries) and combustion (e.g., waste incineration) processes. PCBs were introduced into the environment as commercial mixtures of more than 100 different congeners, with the precise congener composition dependent on the manufacturing process. PCDDs and PCDFs are also composed of hundreds of individual congeners, differing in the number and pattern of chlorine substitution. In general, these compounds have low water solubility, low volatility, and high affinity for sediments and soil. Increasing the number of halogens generally increases environmental persistence and bioaccumulative properties, and non-ortho halogen substitutions generally increase toxicity.

Tissue residues of Ah receptor agonists associated with reduced survival range from 0.00000002 mmol/kg (TCDD) to 0.78 mmol/kg (2,2′,3,3′,5,5′-hexachlorobiphenyl; Johnson et al. 1998; Jarvinen and Ankley 1999). This range encompasses 20 reported values for polyhalogenated dioxin, furan, and biphenyl congeners, and excludes studies reporting only residue levels in eggs or residue levels of total PCBs. Tests were primarily conducted in salmonids, including chinook salmon, coho salmon (*Oncorhynchus kisutch*), rainbow trout, and lake trout (*Salvelinus namaycush*). Other species included carp (*Cyprinus carpio*), guppies, medaka (*Oryzias latipes*), mummichog (*Fundulus heteroclitus*), goldfish, and *Gambusia*. Tissue residues of TCDD that reduced survival ranged from 0.00000002 to 0.0068 mmol/kg (Johnson et al. 1998; Jarvinen and Ankley 1999), a nearly 340,000-fold variation in CBRs for one chemical.

Chemicals with a mechanism of action that requires receptor binding may be extremely species specific and dependent on life stage or environmental modulation because of differences in receptor affinity or number of receptors (Barron et al. 1997). The toxicity of chemicals that bind to the Ah receptor has been shown to depend on Ah receptor expression (Whitlock 1993) such that organisms with the highest concentrations of the Ah receptor will generally be the most sensitive to TCDD toxicity (Willett et al. 2000). In evaluating the Jarvinen and Ankley (1999) database, no clear differences in CBRs were observed between invertebrates and fish, despite speculation that invertebrates are less sensitive to TCDD because of lower concentrations of the Ah receptor. In fish, differences in sensitivity between life stages and species have clearly been shown, such as the brook trout studies of Johnson et al. (1998) and Tietge et al. (1998). Tissue concentrations of TCDD that reduced survival by 50% in swim-up fry were 5–6 pg/g fish in dead swim-up fry compared to 0.5–2 pg/g in live swim-up fry (Johnson et al. 1998). In comparison, 1200 pg/g in adult fish had no effect on survival, growth, gonadal development, and egg production (Tietge et al. 1998). Egg injection studies with TCDD demonstrate that CBRs can exhibit substantial species differences even under nearly identical exposure regimes. For example, TCDD concentrations in fish eggs causing lethality to fry varied by approximately 40 fold between fish species (Elonen et al. 1998). Thus, species and life-stage differences in sensitivity to Ah-receptor agonists may account for much of the range in CBRs.

Mehrle et al. (1988) examined the time dependence and latency period of TCDD toxicity in rainbow trout. Tissue residues of TCDD associated with reduced survival of rainbow trout declined with increasing exposure time both during and after TCDD exposure. After 14 d exposure to TCDD, rainbow trout survival was significantly reduced at 70 nmol/kg (0.8 ng/L TCDD treatment). After 28 d TCDD exposure (0.038 ng/L treatment) and 28 d in clean water, trout survival was significantly reduced at 3 nmol/kg. Reduction in TCDD CBRs with exposure concentration and exposure time demonstrates a latency period for TCDD toxicity. In contrast, the tissue concentrations of PCBs required for mortality of fathead minnows appeared to increase with exposure time between 0 to 12 d (van Wezel et al. 1995). Application of the CBR concept across multiple species of aquatic organisms does not appear to be warranted for Ah agonists because of the extreme variability in CBRs.

VI. Metals

Metals occur both in inorganic forms and as organometallic complexes. The inorganic form of toxic metals can include both the free metal ion and complexed aqueous chemical species. Organometallic complexes (e.g., methylmercury and tributyltin) have very different properties from inorganic metal forms and are evaluated separately in this review.

A. Inorganic Metals

Many metals such as potassium, calcium, sodium, and magnesium are essential elements to aquatic biota and are relatively nontoxic. Toxic metals can be essential in trace amounts, such as copper, nickel, and zinc, or can be nonessential elements such as cadmium, lead, and mercury (Mason and Jenkins 1995). Metals produce toxicity through several mechanisms including ionoregulatory disturbance (McDonald and Wood 1993; Playle 1997), respiratory disturbance (Playle 1997), changes in enzyme activity (Weis and Weis 1992), and cellular damage (Mason and Jenkins 1995). The acute toxicity of metals is dependent on the concentrations at the gill surface, rather than on whole-body or internal tissue concentrations (MacRae et al. 1999). In contrast, the chronic toxicity of metals may be determined by internal concentrations and the propensity for binding cellular molecules such as metallothionein that can sequester metals (Mason and Jenkins 1995).

Many studies on both marine and freshwater invertebrates and fish have shown that the rate of metal accumulation is more important than total metal accumulation. This rate-limiting relationship has been shown for effects on survival (Ahsanullah and Williams 1991; Borgmann et al. 1991; Kraak et al. 1992; Niimi and Kissoon 1994; Absil et al. 1996), growth (Rombough and Garside 1982; Borgmann and Norwood 1997), and reproduction (Spehar 1976). Therefore, acute toxicity is related to the rate of accumulation on the gill or respiratory surface of the organism, whereas chronic toxicity appears to be related to the rate of accumulation in internal tissues or organs. Boese et al. (1999) reported that mortality in *Lumbriculus variegatus* was correlated with whole-body copper accumulation but not with total aqueous copper or calculated cupric ion (Cu^{2+}) in exposure water. However, tissue concentrations of copper were not collected from longer accumulation times, and the effect of uptake rate is unclear.

Borgmann and Norwood (1995) observed an association between copper and zinc uptake and mortality in *Hyalella*. Uptake of metal was rapid followed by internal regulation of the metal, which resulted in a gradual decline of metal concentrations toward preexposure levels after the first 7 d of metal exposure. The measurement of relevant metal residues in aquatic organisms can be confounded by metal binding to exoskeleton and accumulation in the digestive tract (Borgmann and Norwood 1999; Neumann et al. 1999). Despite these caveats, association between toxicity and whole-body residues of certain metals have been shown in some aquatic invertebrates (e.g., *Hyalella* and *Lumbriculus*; Borgmann and Norwood 1995; Boese et al. 1999), and between toxicity and gill tissue residues in some species of fish (Playle et al. 1992, 1993a,b).

Variables that have been shown to influence the tissue residues associated with toxic effects of metals include the exposure route and conditions, intrinsic species differences, and diet before or during exposures. Numerous studies have shown that the aqueous bioavailability of metals affects the degree of toxicity, even though the organisms were exposed to and accumulated similar concentrations of the metal (Hamilton et al. 1990; Bianchini and Gilles 1996; Olge and Knight 1996; Stouthard et al. 1996; Borgmann and Norwood 1997). Different exposure routes such as water or diet can result in tissue concentrations either showing no effects or impaired growth (Besser et al. 1993; Nebeker et al. 1995). Continuous or intermittent exposure regimes have been shown to result in different CBRs (Seim et al. 1984), and closely related species have been shown to accumulate vastly different tissue residues associated with toxic effects (Berlin et al. 1981; Carr et al. 1985). The quality and quantity of food fed before and during an exposure have been shown to influence bioaccumulation as well as residues associated with toxicity in a variety of species (Dixon and Hilton 1985; Meador 1993; Pelgrom et al. 1994; Absil et al. 1996). The influence of these variables on accumulation and toxicity associated with tissue residues greatly limits the application of the CBR concept to most inorganic metals.

Accumulation of metals in fish is relatively slow, and toxicity appears to be related to the rate of metal uptake rather than a critical metal residue (Marr et al. 1996; Hansen et al. 2001). For example, growth reduction in rainbow trout was correlated to exposure time and tissue accumulation of copper in studies using different water hardness and metal exposure concentrations (Marr et al. 1996; Hansen et al. 2001). Arsenate (As^{5+}) CBRs for lethality differed between chronic (0.053 mmol/kg) and acute (0.108–0.115 mmol/kg) exposures (McGeachy and Dixon 1990, 1992). The difference in CBR concentrations may indicate either that there are different mechanisms of mortality for acute and chronic exposures or that the rate of accumulation is important. The relationship between toxic effects and exposure duration diminishes the applicability of the CBR approach for metals in aquatic environments. Overall, the applicability of the CBR concept to inorganic metals appears to be greatly limited. Arsenic CBRs appear to be relatively constant, whereas the CBR approach does not appear to be applicable to copper, cadmium, zinc, mercury, and selenium. Insufficient data were available to evaluate lead, nickel, aluminum, antimony, chromium, and other metals.

B. Organometallic Chemicals

Organometallic chemicals are organic forms of metals that are hydrophobic and can be extensively bioaccumulated. This review focused on important environmental contaminants that have sufficient information for evaluating CBRs: methylmercury and organotin compounds. Methylmercury is produced by bacterial methylation of inorganic mercury (Stumm and Morgan 1981) is highly persistent in the aqueous environment. Methylmercury is soluble in water, but upon entering the organism converts to the methylmercuric ion, which can bind to sulfhydryl groups of proteins in cell membranes (Spacie et al. 1995). Organotins are produced by industrial processes for use as biocides. The most common organotins are tributyltin ($[CH_3(CH_2)_3]_3Sn^+$) which has been widely used in antifouling paints for watercraft, and triphenyltin ($[C_6H_5]_3Sn^+$) which is widely used as a fungicide (Tas et al. 1991, 1996). Both organotin compounds are primarily cationic, moderately hydrophobic, and highly lipophilic.

Methylmercury does not appear to accumulate to a critical concentration in tissues. As with inorganic metals, the rate of accumulation seems much more important than the terminal body residue concentration (Phillips and Buhler 1978; Niimi and Kissoon 1994). Measurements taken in rainbow trout tissues at death revealed that lower mercury concentrations in kidney, liver, spleen, brain, and muscle were found after exposure to higher methylmercury concentrations (Niimi and Kissoon 1994). Therefore, the measured residues at death were inversely proportional to the exposure concentration. The gill was the only tissue that contained similar mercury concentrations at death for all exposures, but the study design did not allow a conclusion of a CBR for gill tissue. The available data suggest that the CBR concept does not apply to methylmercury. However, further research in this area should be conducted to investigate fish gills and organisms other than fish.

Tissue residues of organotin compounds associated with adverse effects appear to be relatively consistent, based on five studies that specifically investigated the CBR concept (Moore et al. 1991; Tas et al. 1991, 1996; Meador 1993, 1997). Meador (1997) reported CBRs for tributyltin between 0.140 and 0.210 mmol/kg for three marine amphipods (*Rhepoxynius abronius, Eohaustorius estuarius,* and *E. washingtonianus*), one marine polychaete (*Armandia brevis*), and one marine flatfish (*Platichthys stellatus*). CBRs for tributyltin in guppies ranged from 0.010 and 0.030 mmol/kg (Tas et al. 1996), and from 0.103 and

0.245 mmol/kg in two amphipods, *R. abronius* and *E. estuarius* (Meador 1993). Moore et al. (1991) reported a lethal tissue residue for tributyltin of 0.062 mmol/kg for dietary exposures of the marine polychaete *Neanthes arenaceodentata*. CBRs for reduced survival span a range of one order of magnitude. Both exposure concentration and internal dose appear to be correlated with specific toxic effects, but between species tissue residue was a much better indicator of toxicity than LC_{50} values (Meador 1997). Accumulation of 0.022 mmol/kg of tributyltin reduced growth and reproduction (Moore et al. 1991), and in two studies CBRs for lethality to triphenyltin ranged from 0.017 to 0.025 mmol/kg (Tas et al. 1991, 1996).

The effect of lipid content on the bioaccumulation of organotin compounds is uncertain. Meador (1993) found that the CBRs for tributyltin mortality were different for amphipods depending on holding time in the laboratory before testing. Measured lipid content in these organisms was also dependent on holding time, and CBRs were more similar when normalized against lipid content. However using a similar study design, Meador (1997) concluded that lipid content was not important, which was attributed to the moderate hydrophobicity and the ionic nature of tributyltin. Overall, these conflicting results introduce considerable uncertainty regarding the importance of lipid content on CBRs for organotin compounds. Evidence for CBRs for triphenyltin effects on survival and tributyltin effects on growth and reproduction is more limited, but are consistent with the CBR concept. Lethal body burdens of tributyltin have been shown to be independent of exposure concentration or exposure duration (Tas et al. 1991), and CBRs for different exposure routes were similar (Moore et al. 1991; Tas et al. 1996; Meador 1997).

VII. Discussion

Environmental applications of the CBR approach will require that the dependence on chemical, biological, and environmental factors is minimal or that the determinants of CBRs can be consistently defined. Previous examinations of the consistency and applicability of CBRs (McCarty and Mackay 1993; Barron et al. 1997) concluded that the primary determinant of a CBR is the mode of action of the chemical. For example, Table 1 shows a 10-million-fold difference in lethal tissue residues for various modes of toxicity, with narcotic chemicals showing the highest CBRs and Ah receptor agonists the lowest CBRs. The current review demonstrated that there is an association between tissue residues and adverse effects in aquatic organisms, but the variability in reported CBRs is extremely large. Expansion of the data reviewed by McCarty and Mackay (1993) to include newer literature and the residue effects database of Jarvinen and Ankley (1999) revealed a 14,000-fold range in CBRs affecting survival for nonpolar narcotics and a 39-million-fold range in CBRs for Ah-receptor agonists (Table 3).

Table 3. Ranges of CBRs affecting survival for organic chemical mode-of-action classes.

Chemical class	CBR range (mmol/kg)	Number of values	Max:min ratio	Data source
Narcotics[a]	9×10^{-3} to 450 (all) NP: 3.2×10^{-2} to 450 polar: 9×10^{-3} to 4.9	144 (all) NP: 80 polar: 64	50,000 (all) NP: 14,000 polar: 540	Jarvinen and Ankley (1999)
Excitatory agents[b]	4×10^{-4} to 0.91	72	2300	Jarvinen and Ankley (1999) Fisher et al. (1999)
AChE inhibitors[c]	4×10^{-5} to 29	115	730,000	Jarvinen and Ankley (1999) Deneer et al. (1999)
Reactive chemicals	9.4×10^{-2} to 13	8	140	Jarvinen and Ankley (1999) McKim and Schmieder (1991)
CNS seizure agents	2×10^{-5} to 1.1 (all) OCs: 7×10^{-5} to 1.1 py[d]: 2×10^{-5} to 1.2×10^{-2}	125 (all) OCs: 106 py[d]: 19	55,000 (all) OCs: 16,000 py[d]: 600	Jarvinen and Ankley (1999)
Ah receptor agonists[e]	2×10^{-8} to 0.78	20	39,000,000	Jarvinen and Ankley (1999) Johnson et al. (1998)

[a]Excludes trichlorophenols and three ring or larger PAHs; NP, nonpolar narcotics; polar, polar narcotics; [b]Includes trichlorophenols; [c]Range, number of values, and max:min ratio exclude one extreme low value (see text for discussion); [d]Separate range, number of values, and max:min ratio determined for organochlorine pesticides (OCs) and pyrethroids (py); [e]Range for TCDD and PCB congeners; excludes studies where only egg residues were reported.

Factors contributing to the variability in CBRs include exposure conditions and chemical structure. The CBRs can also vary substantially in different species and life stages of aquatic organisms, likely because of differences in the disposition of the chemical within the organism, biotransformation, and intrinsic toxicity or inherent sensitivity (Barron et al. 1997; DiToro et al. 2000). The current CBR approach does not incorporate species differences and predicts that CBRs should be relatively constant across taxonomic groups. Some chemicals may accumulate to a greater extent in certain tissues, or may cause toxicity in specific tissues, and metals and reactive chemicals may cause acute toxicity at the gill surface rather than within the organism. Therefore, the CBR approach may not apply to whole-body chemical residue analysis but may be more applicable to specific tissues. Because of the substantial dependence of CBRs on the species, exposure regime, and chemical structure, environmental applications of CBRs do not appear practical at this time.

According to the CBR concept, aquatic organisms with higher lipid content should require higher tissue residues for mortality (McCarty et al. 1992), but lipid content is not directly a part of the CBR model nor is it expressed in the predicted range of CBRs for various chemicals (see Table 1). The target lipid model (DiToro et al. 2000) directly accounts for the lipid content in an exposed organism by expressing the body burden of a chemical in mmol/kg lipid. The reduced variation in narcotic chemical CBRs observed in this model may result from lipid normalization, from accounting for differences in species sensitivity, or from categorizing narcotic chemicals into smaller subclasses that better account for differences in chemical potency. However, lipid normalization may only be applicable to chemicals such as narcotics that have a mechanism of action associated with binding to the lipid components of cells and nerve tissue.

The CBR concept predicts that tissue residues will exhibit a dose–response relationship with adverse effects; that is, adverse effects will increase as tissue residues increase above a threshold concentration of chemical. In general, a dose–response relationship between tissue residues and effects is evident for survival of aquatic organisms exposed to narcotics and some other chemical classes (Figs. 4,5). However, some studies have failed to show a dose–response relationship between tissue residues and adverse effects (Fig. 9), and CBRs for reduced survival can be very similar to no-effect concentrations (see Fig. 2). Nevertheless, a number of individual chemicals appear to accumulate to a CBR that is independent of chemical, biological, and environmental variables, suggesting that tissue residue-based measures of toxicity could offer an appropriate alternative to simple measures of aqueous exposure for some chemicals. However, additional research is needed to confirm CBRs of individual chemicals in controlled laboratory studies, to identify interspecific differences in sensitivity, and to ensure that environmental variables such as temperature and pH do not confound the application of CBRs for individual chemicals.

Overall, it appears that the magnitude of variability of adverse effects of tissue residues, both between chemicals and between taxa, is too large within

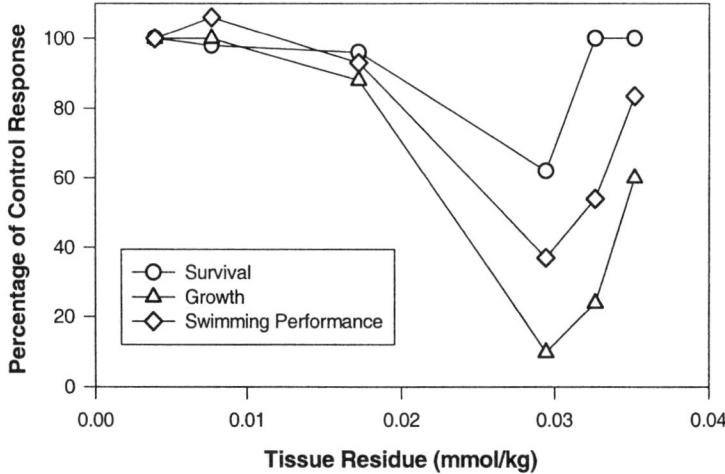

Fig. 9. Dose–response relationships between tissue residues of total naphthalenes and survival, growth, and swimming performance of rainbow trout (*Oncorhynchus mykiss*) exposed to oil for 90 d. (Data from Woodward et al. 1983.)

chemical classes to warrant extensive further investigation and validation of the CBR concept as outlined by McCarty et al. (1993). However, focused QSAR-type approaches such as the target lipid model (DiToro et al. 2000) that incorporate differences in chemical potency, lipid content, and species sensitivity appear to offer potential advances in the applicability of the CBR approach. Additional research is recommended to further reduce variability in predicted versus measured CBRs, focusing on smaller subsets of chemicals that are environmentally important rather than on broad classes of chemicals. Empirical studies should be designed to explicitly test the hypothesis of whether tissue residues can be deemed "critical" rather than assuming that a measured tissue burden is a CBR. This approach requires that laboratory studies be designed to measure tissue residues of both parent compounds and potential metabolites, ideally for whole body and different target organs, across different aqueous exposure conditions (e.g., environmentally relevant pH, hardness/alkalinity, temperature conditions) and for different exposure durations. Adverse effects then should be quantitatively related to tissue levels using dose–response models. Further, testing should be performed for several species to identify interspecific differences in sensitivity.

Summary

Associations between tissue residues and toxicity to aquatic organisms were examined to evaluate the applicability of the critical body residue (CBR) approach across different chemical classes. Chemical classes and mode of action categories evaluated included narcotics (polar and nonpolar), excitatory agents, AChE inhibitors, reactives/irritants, CNS seizure agents, aryl hydrocarbon (Ah) receptor agonists, and inorganic metals and organometals. This evaluation indicated that empirical data do not support broad application of the CBR concept across chemical classes. This conclusion is particularly important for polar and nonpolar narcotics because the CBR concept was specifically developed for these chemical classes. The variability observed in tissue residues between chemicals within a given mode-of-action class appears to be generally of the same order of magnitude as the variability of aqueous measures of toxicity such as LC_{50} values (Table 3; Fig. 10). This observation suggests that either (a) the reported tissue residues were dependent on the aqueous dosing regime; (b) the tissue measurements do not accurately reflect the internal dose at target organs with substantially greater precision than water exposure measurements; or (c) many of the same sources of variability associated with aqueous exposures, such as chemical structure, individual species sensitivity, biotransformation processes, and lipid content, also apply to tissue-based measures of exposure. An additional source of uncertainty of CBRs is whether a chemical has been correctly assigned to a mode of action category.

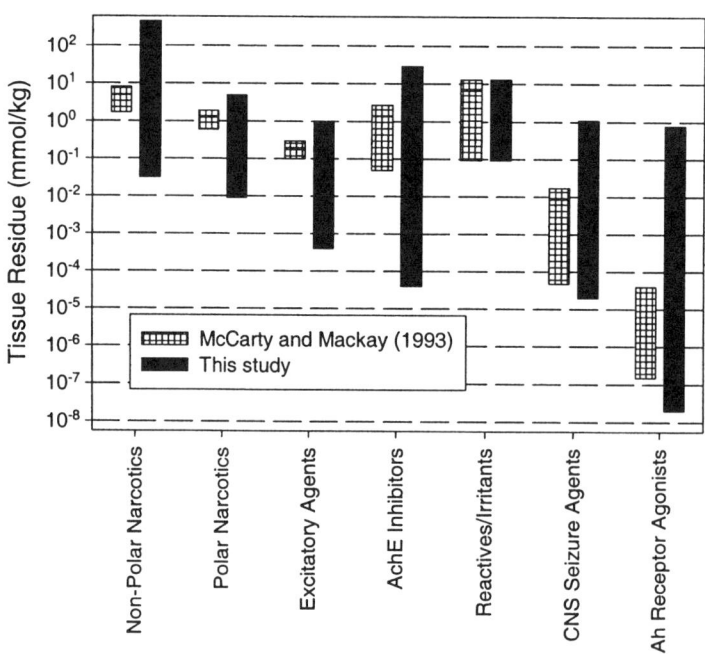

Fig. 10. Comparison of CBRs reported by McCarty and Mackay (1993) with CBR ranges determined in the current study.

The CBR approach outlined by McCarty (1986, 1987) and McCarty et al. (1993) underlines an important concept in aquatic toxicology, i.e., that internal chemical dose is the true measure of toxicity for many chemicals rather than imputed dose based on aqueous exposure. Nevertheless, without more refined and accurate examination of that actual internal dose and without additional consideration of differences in sensitivity between species, differences in toxic potency between chemicals, and differences in toxicity of environmentally modified or biotransformed compounds, the CBR approach may not offer practical advantages over conventional media-based exposure assessment.

Acknowledgments

We thank Jim Clark and Don Versteeg for review and comments. This study was funded by the American Chemistry Council.

References

Absil MCP, Berntssen M, Gerringa LJA (1996) The influence of sediment, food and organic ligands on the uptake of copper by sediment-dwelling bivalves. Aquat Toxicol 34:13–29.

Ahsanullah M, Williams AR (1991) Sublethal effects and bioaccumulation of cadmium, chromium, copper, and zinc in the marine amphipod *Allorchestes compressa*. Mar Biol 108:59–65.

Ankley GT, Erickson RJ, Phipps GL, Mattison VR, Kosian PA, Sheedy BR, Cox JS (1995) Effects of light intensity on the phototoxicity of fluoranthene to a benthic macroinvertebrate. Environ Sci Technol 29:2828–2833.

Arthur AD (1991) Verification studies of body residue-based model for predicting the sublethal toxicity of time variable exposures to organic chemicals in small fish. Masters thesis, University of Waterloo, Ontario, Canada.

Barron MG, Woodburn KB (1995) Ecotoxicology of chlorpyrifos. Rev Environ Contam Toxicol 144:1–93.

Barron MG, Anderson MJ, Lipton J, Dixon DG (1997) Evaluation of critical body residue QSARs for predicting organic chemical toxicity to aquatic organisms. SAR QSAR Environ Res 6:47–62.

Barron MG, Little EE, Calfee RD, Diamond S (2000) Quantifying solar spectral irradiance in aquatic habitats for the assessment of photoenhanced toxicity. Environ Toxicol Chem 19:920–925.

Berlin WH, Hasselberg RJ, Mac MJ (1981) Growth and mortality of fry of Lake Michigan lake trout during chronic exposure to PCBs and DDE in chlorinated hydrocarbons as a factor in the reproduction and survival of Lake Trout (*Salvelinus namaycush*) in Lake Michigan. Technical paper 105. U.S. Fish and Wildlife Service, Ann Arbor, MI, pp 11–22.

Besser JM, Canfield TJ, LaPoint TW (1993) Bioaccumulation of organic and inorganic selenium in a laboratory food chain. Environ Toxicol Chem 12:57–72.

Bianchini A, Gilles R (1996) Toxicity and accumulation of mercury in three species of crabs with different osmoregulatory capacities. Bull Environ Contam Toxicol 57: 91–98.

Boese CJ, Collyard SA, Bergman HL, Meyer JS (1999) Whole-body copper accumulation predicts acute toxicity to *Lumbriculus variegatus* as pH and calcium vary. Presentation at the Annual Meeting of the Society of Environmental Toxicology and Chemistry, Philadelphia, PA (abstract in proceedings), p 239.

Borgmann U, Norwood WP (1995) Kinetics of excess (above background) copper and zinc in *Hyalella azteca* and their relationship to chronic toxicity. Can J Fish Aquat Sci 52:864–874.

Borgmann U, Norwood WP (1997) Toxicity and accumulation of zinc and copper in *Hyalella azteca* exposed to metal-spiked sediments. Can J Fish Aquat Sci 54:1046–1054.

Borgmann U, Norwood WP (1999) Assessing the toxicity of lead in sediments to *Hyalella azteca*: the significance of bioaccumulation and dissolved metal. Can J Fish Aquat Sci 56:1494–1503.

Borgmann U, Norwood WP, Babirad IM (1991) Relationship between chronic toxicity and bioaccumulation of cadmium in *Hyalella azteca*. Can J Fish Aquat Sci 48:1055–1060.

Bradbury SP, Symonik DM, Coats JR, Atchison GJ (1987) Toxicity of fenvalerate and its constituent isomers to the fathead minnow, *Pimephales promelas*, and bluegill, *Lepomis macrochirus*. Bull Environ Contam Toxicol 38:727–735.

Bradbury SP, Henry TR, Niemi GJ, Carlson RW, Snarski VM (1989) Use of respiratory-cardiovascular responses of rainbow trout (*Oncorhynchus mykiss*) in identifying acute toxicity syndromes in fish: Part 3. Polar narcotics. Environ Toxicol Chem 8:247–261.

Carr RS, Williams JW, Saksa FI, Buhl RL, Neff JM (1985) Bioenergetic alterations correlated with growth, fecundity and body burden of cadmium for mysids (*Mysidopsis bahia*). Environ Toxicol Chem 4:181–188.

Connell D, Markwell R (1992) Mechanism and prediction of nonspecific toxicity to fish using bioconcentration characteristics. Ecotoxicol Environ Saf 24:247–265.

de Maagd PG-J, van de Klundert ICM, van Wezel AP, Opperhuizen A, Sijm DTHM (1997) Lipid content and time-to-death-dependent lethal body burden of napthalene and 1,2,4-trichlorobenzene in fathead minnow (*Pimephales promelas*). Ecotoxicol Environ Saf 38:232–237.

Deneer JW, Budde BJ, Weijers A (1999) Variations in the lethal body burdens of organophosphorus compounds in the guppy. Chemosphere 38:1671–1683.

DiToro DM, McGrath JA (2000) Technical basis for narcotic chemicals and PAH criteria: II. Mixtures and sediments. Environ Toxicol Chem 19:1971–1982.

DiToro DM, McGrath JA, Hansen DJ (2000) Technical basis for narcotic chemicals and PAH criteria: I. Water and tissue. Environ Toxicol Chem 16:1951–1970.

Dixon DG, Hilton JW (1985) Effects of available dietary carbohydrate and water temperature on the chronic toxicity of waterborne copper to rainbow trout (*Salmo gairdneri*). Can J Fish Aquat Sci 42:1007–1013.

Elonen GE, Spehar RL, Holcombe GW, Johnson RD, Fernandez JD, Erickson RJ, Tietge JE, Cook PM (1998) Comparative toxicity of 2,3,7,8-tetrachlorodibenzo-*p*-dioxin to seven freshwater fish species during early life-stage development. Environ Toxicol Chem 17:472–483.

Fisher SW, Hwang H, Atanasoff M, Landrum PF (1999) Lethal body residues for pentachlorophenol in zebra mussels (*Dreissena polymorpha*) under varying conditions of temperature and pH. Ecotoxicol Environ Saf 43:274–283.

Hamilton SJ, Buhl KJ, Faerber NL, Wiedmeyer RH, Bullard FA (1990) Toxicity of organic selenium in the diet of chinook salmon. Environ Toxicol Chem 9:347–358.

Hansen JA, Lipton J, Welsh P, Morris J, Cacela D, Suedkamp M (2001) Relationship between water exposure, tissue residues, growth, and mortality of rainbow trout (*Oncorhynchus mykiss*) exposed to copper. Aquat Toxicol (in press).

Hermens JLM (1990) Electrophiles and acute toxicity to fish. Environ Health Perspect 87:219–225.

Hickie BE, McCarty LS, Dixon DG (1995) A residue-based toxicokinetic model for pulse-exposure toxicity in aquatic systems. Environ Toxicol Chem 14:2187–2197.

Howe GE, Marking LL, Bills TD, Rach JJ, Mayer FL (1994) Effects of water temperature and pH on toxicity of terbufos, trichlorfon, 4-nitrophenol and 2,4-dinitrophenol to the amphipod *Gammarus pseudolimnaeus* and rainbow trout (*Oncorhynchus mykiss*). Environ Toxicol Chem 13:51–66.

Jarvinen AW, Ankley GT (1999) Linkage of Effects to Tissue Residues: Development of a Comprehensive Database for Aquatic Organisms Exposed to Inorganic and Organic Chemicals. SETAC Press, Pensacola, FL.

Johnson RD, Tietge JE, Jensen KM, Fernandez JD, Linnum AL, Lothenbach DB, Holcombe GW, Cook PM, Christ SA, Lattier DL, Gordon DA (1998) Toxicity of 2,3,7,8-tetrachlorodibenzo-*p*-dioxin to early life stage brook trout (*Salvelinus fontinalis*) following parental dietary exposure. Environ Toxicol Chem 17:2408–2421.

Kane Driscoll SB, Schaffner LC, Dickhut RM (1998) Toxicokinetics of fluoranthene to the amphipod, *Leptocheirus plumulosus*, in water-only and sediment exposures. Mar Environ Res 45:269–284.

Kishino T, Kobayashi K (1995) Relationship between toxicity and accumulation of chloropenols at various pH, and their absorption mechanism in fish. Water Res 29:431–442.

Klee U (1998) Evaluation of a body-residue-based one-compartment first-order kinetic model for estimating the toxicity of mixtures of chlorinated organic contaminants to rainbow trout (*Oncorhynchus mykiss*). PhD thesis, University of Waterloo, Ontario, Canada.

Kraak MHS, Lavy D, Peeters WHM, Davids C (1992) Chronic ecotoxicity of copper and cadmium to the zebra mussel *Dreissena polymorpha*. Arch Environ Contam Toxicol 23:363–369.

Landrum PF, Dupuis DW (1990) Toxicity and toxicokinetics of pentachlorophenol and carbaryl to *Pontoporeia hoyi* and *Mysis relicta*. In: Landis WG, van der Schalie WH (eds) Aquatic Toxicology and Risk Assessment, Vol. 13. American Society for Testing and Materials (ASTM), Philadelphia, PA, pp 278–289.

Lee J-H, Landrum PF, Koh C-H (1999) Change in critical body residues of PAH in *Hyalella azteca* during 10-day water-only exposure period. Presentation at the Annual Meeting of the Society of Environmental Toxicology and Chemistry, Philadelphia, PA (abstract in proceedings), p 296.

Legierse KCHM, Verhaar HJM, Vaes WHJ, de Bruijn JHM, JLM Hermens (1999) Analysis of time-dependent acute aquatic toxicity of organophosphorus pesticides: the critical target occupation model. Environ Sci Technol 33:917–925.

MacRae RK, Smith DE, Swoboda-Colberg N, Meyer JS, Bergman HL (1999) Copper binding affinity of rainbow trout (*Oncorhynchus mykiss*) and brook trout (*Salvelinus fontinalis*) gills: implications for assessing bioavailable metal. Environ Toxicol Chem 18:1180–1189.

Marr JCA, Lipton J, Cacela D, Hansen JA, Bergman HL, Meyer JS, Hogstrand C (1996) Relationship between copper exposure duration, tissue copper concentration, and rainbow trout growth. Aquat Toxicol 36:17–30.

Mason AZ, Jenkins KD (1995) Metal detoxification in aquatic organisms. In: Tessier A, Turner DR (eds) Metal Speciation and Bioavailability in Aquatic Systems, Wiley, Chichester, pp 479–608.

McCarty LS (1986) The relationship between aquatic toxicity QSARs and bioconcentration for some organic chemicals. Environ Toxicol Chem 5:1071–1080.

McCarty LS (1987) Relationship between toxicity and bioconcentration for some organic chemicals. 1. Examination of the relationship. In: Kaiser KLE (ed) QSAR in Environmental Toxicology, Vol. II. Reidel, Dordrecht, pp 207–220.

McCarty LS, Mackay D (1993) Enhancing ecotoxicological modeling and assessment. Environ Sci Technol 27:1719–1728.

McCarty LS, Mackay D, Smith AD, Ozburn GW, Dixon DG (1992) Residue-based interpretation of toxicity and bioconcentration QSARs from aquatic bioassays: neutral narcotic organics. Environ Toxicol Chem 11:917–930.

McCarty LS, Mackay D, Smith AD, Ozburn GW, Dixon DG (1993) Residue-based interpretation of toxicity and bioconcentration QSARs from aquatic bioassays: polar narcotic organics. Ecotoxicol Environ Saf 25:253–270.

McDonald DG, Wood CM (1993) Branchial mechanisms of acclimation to metals in freshwater fish. In: Rankin JC, Jensen FB (eds) Fish Ecophysiology. Chapman & Hall, London, pp 297–321.

McGeachy SM, Dixon DG (1990) Effect of temperature on the chronic toxicity of arsenate to rainbow trout (*Oncorhynchus mykiss*). Can J Fish Aquat Sci 47:2228–2234.

McGeachy SM, Dixon DG (1992) Whole-body arsenic concentrations in rainbow trout during acute exposure to arsenate. Ecotoxicol Environ Saf 24:301–308.

McKim JM, Schmieder PK (1991) Bioaccumulation: does it reflect toxicity? In: Nagel R, Loskill R (eds) Bioaccumulation in Aquatic Systems. VCH Verlagsgesellschaft mbH, Weinheim, Germany, pp 161–188.

Meador JP (1993) The effect of laboratory holding on the toxicity response of marine infaunal amphipods to cadmium and tributyltin. J Exp Mar Biol Ecol 174:227–242.

Meador JP (1997) Comparative toxicokinetics of tributyltin in five marine species and its utility in predicting bioaccumulation and acute toxicity. Aquat Toxicol 37:307–326.

Mehrle PM, Buckler DR, Little EE, Smith LM, Petty JD, Peterman PH, Stalling DL, DeGraeve DG, Cole JJ, Adams WJ (1988) Toxicity and bioconcentration of 2,3,7,8-tetrachlorodibenzodioxin and 2,3,7,8-tetrachlorodibenzofuran in rainbow trout. Environ Toxicol Chem 7:47–62.

Moore DW, Dillon TM, Suedel BC (1991) Chronic toxicity of tributyltin to the marine polychaete worm, *Neanthes arenaceodentata*. Aquat Toxicol 21:181–198.

Nebeker AV, Schuytema GS, Ott SL (1995) Effects of cadmium on growth and bioaccumulation in the northwestern salamander *Ambystoma gracile*. Arch Environ Contam Toxicol 29:492–499.

Neumann PTM, Borgmann U, Norwood W (1999) Effect of gut clearance on metal body concentrations in *Hyalella azteca*. Environ Toxicol Chem 18:976–984.

Niimi AJ, Kissoon GP (1994) Evaluation of the critical body burden concept based on inorganic and organic mercury toxicity to rainbow trout (*Oncorhynchus mykiss*). Arch Environ Contam Toxicol 26:169–178.

Olge RS, Knight AW (1996) Selenium bioaccumulation in aquatic ecosystems: 1. Effects of sulfate on the uptake and toxicity of selenate in *Daphnia magna*. Arch Environ Contam Toxicol 30:274–279.

Pawlisz AV, Peters RH (1993a) A radiotracer technique for the study of lethal body burdens of narcotic organic chemicals in *Daphnia magna*. Environ Sci Technol 27: 2795–2800.

Pawlisz AV, Peters RH (1993b) A test of the equipotency of internal body burdens of nine narcotic chemicals using *Daphnia magna*. Environ Sci Technol 27:2801–2806.

Pelgrom SMGJ, Lamers LPM, Garritsen JAM, Pels BM, Lock RAC, Balm PHM, Wendelaar Bonga SE (1994) Interactions between copper and cadmium during single and combined exposure in juvenile tilapia *Oreochromis mossambicus*: influence of feeding condition on whole body metal accumulation and the effect of the metals on tissue water and ion content. Aquat Toxicol 30:117–135.

Penttinen O-P, Kukkonen J (1998) Chemical stress and metabolic rate in aquatic invertebrates: threshold, dose-response relationships, and mode of toxic action. Environ Toxicol Chem 17:883–890.

Phillips GR, Buhler DR (1978) The relative contributions of methylmercury from food or water to rainbow trout (*Salmo gairdneri*) in a controlled laboratory environment. Trans Am Fish Soc 107:853–861.

Playle RC (1997) Physiological and toxicological effects of metals at gills of freshwater fish. In: Bergman HL, Dorward-King EJ (eds) Reassessment of Metals Criteria for Aquatic Life Protection. SETAC Press, Pensacola, FL, pp 101–105.

Playle RC, Gensemer RW, Dixon DG (1992) Copper accumulation on gills of fathead minnows: influence of water hardness, complexation and pH of the gill micro-environment. Environ Toxicol Chem 11:381–391.

Playle RC, Dixon DG, Burnison K (1993a) Copper and cadmium binding to fish gills: estimates of metal-gill stability constants and modelling of metal accumulation. Can J Fish Aquat Sci 50:2678–2687.

Playle RC, Dixon DG, Burnison K (1993b) Copper and cadmium binding to fish gills: modification by dissolved organic carbon and synthetic ligands. Can J Fish Aquat Sci 50:2667–2677.

Rombough PJ, Garside ET (1982) Cadmium toxicity and accumulation in eggs and alevins of Atlantic salmon *Salmo salar*. Can J Zool 60:2006–2014.

Russom CL, Bradbury SP, Broderius SJ, Hammermeister DE, Drummond RA (1997) Predicting modes of toxic action from chemical structure: acute toxicity in the fathead minnow (*Pimephales promelas*). Environ Toxicol Chem 16:948–967.

Seim WK, Curtis LR, Glenn SW, Chapman GA (1984) Growth and survival of developing steelhead trout (*Salmo gairdneri*) continuously or intermittently exposed to copper. Can J Fish Aquat Sci 41:433–438.

Sijm DTHM, Schipper M, Opperhuizen A (1993) Toxicokinetics of halogenated benzenes in fish: lethal body burden as a toxicological endpoint. Environ Toxicol Chem 12:1117–1127.

Spacie A, McCarty LS, Rand GM (1995) Bioaccumulation and bioavailability in multiphase systems. In: Rand GM (ed) Fundamentals of Aquatic Toxicology: Effects, Environmental Fate, and Risk Assessment, 2nd Ed. Taylor & Francis, Washington, DC, pp 493–521.

Spehar RL (1976) Cadmium and zinc toxicity to flagfish (*Jordanella floridae*). J Fish Res Board Can 33:1939–1945.

Stouthard XJHX, Haans JLM, Lock RAC, Wendelaar Bonga SE (1996) Effects of water pH on copper toxicity to early life stages of the common carp (*Cyprinus carpio*). Environ Toxicol Chem 15:376–383.

Stumm W, Morgan JJ (1981) Aquatic Chemistry: An Introduction Emphasizing Chemical Equilibria in Natural Waters. Wiley, New York.

Tas JW, Seinen W, Opperhuizen A (1991) Lethal body burden of triphenyltin chloride in fish: preliminary results. Comp Biochem Physiol 100C:59–60.

Tas JW, Keizer A, Opperhuizen A (1996) Bioaccumulation and lethal body burden of four triorganotin compounds. Bull Environ Contam Toxicol 57:146–154.

Tietge JE, Johnson RD, Jensen KM, Cook PM, Elonen GE, Fernandez JD, Holcombe GW, Lothenbach DB, Nichols JW (1998) Reproductive toxicity and disposition of 2,3,7,8-tetrachlorodibenzo-*p*-dioxin in adult brook trout (*Salvelinus fontinalis*) following a dietary exposure. Environ Toxicol Chem 17:2395–2407.

Vaes WHJ, Ramos EU, Verhaar HJM, Hermens JLM (1998) Acute toxicity of nonpolar versus polar narcosis: is there a difference? Environ Toxicol Chem 17:1380–1384.

van Hoogen G, Opperhuizen A (1988) Toxicokinetics of chlorobenzenes in fish. Environ Toxicol Chem 7:213–219.

van Wezel AP, Jonker MTO (1998) Use of lethal body burden in the risk quantification of field sediments: influence of temperature and salinity. Aquat Toxicol 42:287–300.

van Wezel AP, Opperhuizen A (1995) Narcosis due to environmental pollutants in aquatic organisms: residue-based toxicity, mechanisms, and membrane burdens. Crit Rev Toxicol 25:255–279.

van Wezel AP, de Vries DAM, Kostense S, Sijm DTHM, Opperhuizen A (1995) Intraspecies variation in lethal body burdens of narcotic compounds. Aquat Toxicol 33: 325–342.

van Wezel AP, de Vries DAM, Sijm DTHM, Opperhuizen A (1996) Use of the lethal body burden in the evaluation of mixture toxicity. Ecotoxicol Environ Saf 35:236–241.

Verhaar HJM, de Wolf W, Dyer S, Legierse KCHM, Seinen W, Hermens JLM (1999) An LC_{50} vs time model for the aquatic toxicity of relative and receptor-mediated compounds. Consequences for bioconcentration kinetics and risk assessment. Environ Sci Technol 33:758–763.

Weis P, Weis JS (1992) Metal toxicology and fish development. In: Deposition and Fate of Trace Metals in Our Environment. Symposium Proceedings. North Central Forest Experimental Station, U.S. Department of Agriculture Forest Service, St. Paul, MN, pp 157–171.

Whitlock JP (1993) Mechanistic aspects of dioxin action. Chem Res Toxicol 6:754–763.

Willett KL, Wilson C, Thomsen J, Porter W (2000) Evidence for and against the presence of polynuclear aromatic hydrocarbon and 2,3,7,8-tetrachloro-*p*-dioxin binding proteins in the marine mussels *Bathymodiolus and Modiolus modiolus*. Aquat Toxicol 48:51–64.

Woodward DF, Riley RG, Smith CE (1983) Accumulation, sublethal effects and safe concentrations of a refined oil as evaluated with cutthroat trout. Arch Environ Contamin Toxicol 12:455–464.

Manuscript received August 23, 2000; accepted January 17, 2001.

Toxicity of Azaarenes

Eric A.J. Bleeker, Saskia Wiegman, Pim de Voogt, Michiel Kraak, Heather A. Leslie, Elske de Haas, and Wim Admiraal

Contents

I. Introduction	40
II. Biotransformation and Metabolism	41
A. Introduction	41
B. Quinoline and Isoquinoline	42
C. Benzoquinolines	44
D. Azaarenes with More Than Three Condensed Aromatic Rings	46
E. Metabolites Occurring in the Environment	47
III. Direct Toxicity	48
A. Introduction	48
B. Toxicity of Quinoline and Acridine	49
C. Toxicity of Ranges of Azaarenes	50
D. Biotransformation and Toxicity	52
E. Chronic Toxicity	52
IV. Photochemical Transformation and Phototoxic Effects of Azaarenes	53
A. Introduction	53
B. Mechanisms and Kinetics of Photochemical Reactions of (N)PAHs	53
C. Phototoxic Effects of (N)PAHs and Their Photoproducts on Aquatic Organisms	57
V. Genotoxicity and Carcinogenicity	62
A. Introduction	62
B. Quinoline and Isoquinoline	63
C. Benzoquinolines	63
D. Benzacridines	65
E. Dibenzacridines	66
VI. Teratogenicity and Other Developmental Effects	66
VII. Comparing Azaarenes and Homocyclic PAHs	67
A. Comparing Toxicity	67
B. Risk Assessment	69
VIII. Conclusions	69
Summary	70
References	72

Communicating editor: Pim de Voogt.

E.A.J. Bleeker (✉)·S. Wiegman·M. Kraak·H.A. Leslie·E. de Haas·W. Admiraal
Department of Aquatic Ecology and Ecotoxicology, IBED, University of Amsterdam, Kruislaan 320, 1098 SM Amsterdam, The Netherlands

P. de Voogt
Department of Environmental and Toxicological Chemistry, IBED, University of Amsterdam, Nieuwe Achtergracht 166, 1018 WV Amsterdam, The Netherlands

I. Introduction

Polycyclic aromatic hydrocarbons (PAHs) are molecules composed of two or more fused aromatic rings, either benzene or cyclopentadiene rings or both. They are found in fossil fuels, but apart from oil spills the main anthropogenic sources of PAHs are all types of (oil-related) industry as well as wood preservation and combustion (Khalili et al. 1995; Slooff et al. 1989). Their significance in environmental pollution has been recognized widely since the late 1970s and early 1980s. Since then, many studies have considered their occurrence (Baek et al. 1991; Wilcke 2000), their biotransformation (Ashok and Saxena 1995; Muncnerova and Augustin 1994), and the different types of toxicity they induce, for instance carcinogenicity (Santodonato 1997). PAH emissions often contain substituted compounds such as oxygenated, hydroxylated, chlorinated, nitro- or amino PAHs, each with their specific occurrence, metabolism, and biological impact (Howard et al. 1991).

Apart from on-ring substitutions, a variety of atoms can be incorporated in the aromatic rings, as for example sulfur and nitrogen atoms (Newsted and Giesy 1987). Although these heterocyclic PAHs outnumber the nonsubstituted homocycles (Kuhn and Suflita 1989), they receive relatively little attention; this may be because of the low concentrations of N-, O-, and S-PAHs in the field, which appear to be 1%–10% of those of the analogue homocyclic PAHs (Nielsen et al. 1997). The generally higher water solubility of these heterocyclics (Pearlman et al. 1984), however, may imply a higher environmental impact. This review focuses, therefore, on the role of in-ring substitutions on the environmental fate and toxicity of PAHs.

S-Heterocyclics are relatively nonpolar, suggesting that the environmental impact of this group of compounds is similar to that of homocyclic PAHs. Furthermore, effects of S-heterocyclics have already been reviewed (Kropp and Fedorak 1998). O-heterocyclics are more water soluble, but in the field these compounds are quite reactive, resulting in on-ring substituted compounds; this obviously obscures the possible effects of in-ring substitutions. This review focuses, therefore, on one group of N-heterocycles, the azaarenes, in search for similarities and differences in toxicity and transformation of heterocyclic and homocyclic PAHs.

These azaarenes contain one nitrogen atom in place of a carbon atom (Fig. 1). Apart from their natural origin (e.g., as alkaloids: Kaiser et al. 1996; Michael 2000), azaarenes are formed and released into the environment from sources similar to those of homocyclic PAHs, such as incomplete combustion of fossil fuels, spills or effluents of several industrial activities, and oil drilling, refining, and storage (Kochany and Maguire 1994), coal tar distillation (Pereira et al. 1983), and wood preservation (Adams and Giam 1984; Pereira et al. 1983). In addition, the occurrence of basic azaarene structures as a moiety of pharmaceuticals (Oshiro et al. 1998; Siim et al. 2000) and pesticides (Kuhn and Suflita 1989) increases the need to consider sources of NPAH totally unrelated to the sources of PAH. The emission of PAHs, including azaarenes, into the atmo-

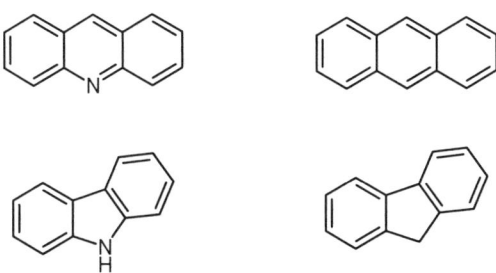

Fig. 1. Acridine and carbazole (left) and their homocyclic analogues anthracene and fluorene (right).

sphere has increased greatly during the past century. Historical records of PAHs in soil and sediment from rural areas and in ice from sites as remote as Greenland document the widespread environmental contamination that has occurred (Jones et al. 1989; Kawamura et al. 1994; Sanders et al. 1993). Blumer et al. (1977), Furlong and Carpenter (1982), Wakeham (1979), and Bleeker et al. (1996) detected elevated azaarene concentrations in marine and freshwater sediments. In addition, Van Genderen et al. (1994) and Kozin et al. (1997) demonstrated the occurrence of azaarenes in freshwaters in the Netherlands.

Knowledge of azaarene toxicity and the role of metabolism in toxicity is fragmentary. This review aims, therefore, to analyze and summarize current knowledge on azaarene metabolism and toxicity, as well as comparing these data with those of homocyclic PAHs. The first part focuses on biotransformation of azaarenes and possible implications for toxicity. The second part is concerned with the different aspects of azaarene toxicity: direct toxicity, photoenhanced toxicity, genotoxicity, carcinogenicity, and teratogenicity and other developmental effects.

II. Biotransformation and Metabolism
A. Introduction

PAHs are subjected to chemical transformation and degradation through a variety of processes. In the aquatic environment, photooxidation, chemical oxidation, and biological transformation by bacteria, fungi, algae, and animals are of primary importance. Abiotic PAH transformation has been reviewed by Kochany and Maguire (1994), and much research has also been done on biotransformation by bacteria (Bollag and Kaiser 1991; Kuhn and Suflita 1989), fungi (Muncnerova and Augustin 1994; Sutherland et al. 1998), algae (Dijkman et al. 1997; Warshawsky et al. 1995), and animals, both vertebrates (McMurtrey and Knight 1984; Warshawsky et al. 1996) and invertebrates (Lee 1988; Livingstone and Farrar 1984).

Biotransformation of chemicals such as azaarenes is necessary for organisms to cope with the thousands of different natural toxicants and xenobiotics that may accumulate and disturb their normal functions. It occurs when the metabolic machinery, usually activated through their presence, processes the xenobiotic, usually attaching a polar group to award a higher degree of water solubility to the chemical to facilitate excretion. The desired result of biotransformation is inactivation and detoxification of xenobiotics. However, during the course of biotransformation processes hazardous reactive intermediates or toxicologically active stable transformation products may be formed, in which case activation (toxification) of the xenobiotic is achieved (see Sections III–VI).

Azaarenes comprise pyrrol and pyridine derivatives. For pyrrol derivatives, quite a few reports on microbial transformations exist (e.g., indole: Kamath and Vaidyanathan 1990; carbazole: Gieg et al. 1996). For methylquinolines, which are usually hydroxylated to their 2-hydroxyquinoline analogues (Johansen et al. 1997a), such reports are also available (Aislabie et al. 1990; Liu et al. 1994b; Sutton et al. 1996). This review is, therefore, confined to the pyridine types of nonsubstituted azaarenes. When such derivatives contain four or more condensed rings, the structural resemblance with their homocyclic analogues is very strong. As a result, their transformations in mammals proceed analogous to homocyclic pathways with hydroxylated and dihydrodiol transformation products (Section V). Therefore, and because we intend to focus on the azaarene-specific transformation, two- and three-ringed structures are the main topic of this section.

Biodegradation occurs both under aerobic and anaerobic conditions. Usually, the initial step involves insertion of an oxygen atom in some way. The oxygen may originate from reactive oxygen species as well as from water, and both mono- and dioxygenases can be involved in particular transformation steps.

Species-specific differences in biotransformation rate are well known. For microorganisms, this specificity appears to be associated with type of bacteria or fungi (Kaiser et al. 1996) and endogenous levels of metabolizing enzymes (Livingstone 1998). The type of substituents also influences transformation rates (Kaiser et al. 1996). Transformation of multiring azaarenes by invertebrates or microorganisms proceeds slowly, if at all, and few metabolites have been identified (Grosser et al. 1995; Siddiqi et al. 1994). From the scarcely documented metabolism, it appears that biotransformation products differ between taxa. In toxicity tests with algae and midge larvae, ketonic products were identified as major transformation products (De Voogt et al. 1999). In vertebrates, dihydrodiols appear to be major metabolites. Fungi transform azaarenes into hydroxylated metabolites as well as dihydrodiols (Sutherland et al. 1994a). Metabolism of quinoline by fish results in hydroxylated and thioazaarenes (Bean et al. 1985).

B. Quinoline and Isoquinoline

Several organisms appear to be able to degrade quinolines. Organisms studied include bacteria, fungi, and fish. Microbial degradation of quinolines generally proceeds via the hydroxylation at the 2-position resulting in the tautomeric forms 2-hydroxyquinoline and 2(1*H*)-quinolinone (Fig. 2). 2(1*H*)-Quinolinone

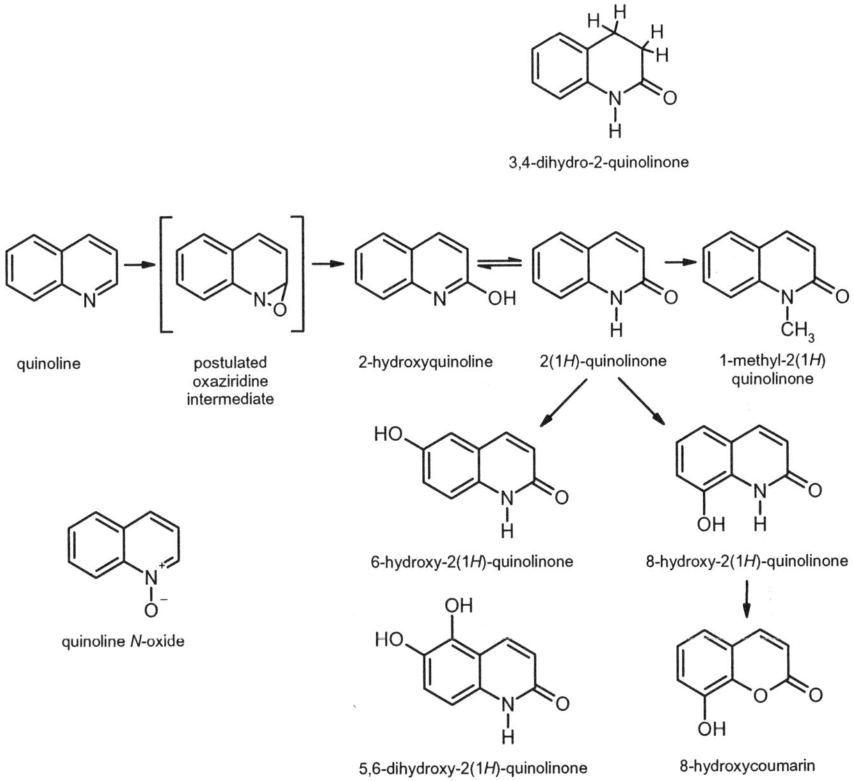

Fig. 2. Microbial degradation of quinoline (as proposed by Boyd et al. 1993; Kilbane et al. 2000; Pereira et al. 1987; Shukla 1986).

is further degraded by microorganisms into either hydroxylated coumarins (Kilbane et al. 2000; Shukla 1986) or stereoisomeric dihydrodiols (Boyd et al. 1993). The coumarin pathway has been confirmed in degradation experiments with *Pseudomonas stutzeri, P. ayucida,* and *P. putida* (Boyd et al. 1993; Kilbane et al. 2000). Hydroxylation at position 2 is also the initial step in the degradation of quinoline (and several methylquinolines) by *Desulfobacterium indolicum.* One of the transformation products was identified as the 3,4-dihydro-2-quinolinone metabolite (Johansen et al. 1997b). Under anaerobic conditions, quinoline is transformed into 2-hydroxyquinoline (Liu et al. 1994a).

Hydroxyquinolines were also found in fish (*Salmo gairdneri*) as transformation products of quinoline (Bean et al. 1985), although in this case hydroxylation appeared at the 3-, 6-, and 7-positions. Furthermore, four *S*-acetylquinolinethiols were found as major transformation products. In experiments with fungi (*Cunninghamella elegans*), quinoline-*N*-oxide (see Fig. 2) was identified as a major metabolite (Sutherland et al. 1994b). The same product has also been found in rat and rabbit liver microsomes when treated with quinoline (Cowan et al. 1978; LaVoie et al. 1983; Tada et al. 1982).

Isoquinoline appears to be degraded more slowly than quinoline, but generally follows analogous breakdown pathways. Thus, anaerobic degradation results in hydroxylation *ortho* to the nitrogen atom followed by tautomerization into the isoquinolinone and further methylation (Fig. 3) (Pereira et al. 1987).

Isoquinoline products from bacterial tests that have been identified include isoquinoline-*N*-oxide, 1-, 4-, 5-, and 8-hydroxyisoquinoline, *cis*-7,8-dihydroisoquinoline-7,8-diol, and *cis*-5,6-dihydroisoquinoline-5,6-diol (Boyd et al. 1993; Pereira et al. 1983). The 1-hydroxyisoquinoline product has also been suggested as an intermediate in aerobic microbial degradation of isoquinoline into CO_2, although no firm confirmation of the identity was provided (Aislabie et al. 1989). Fungi also appear to handle isoquinoline in a similar way as quinoline, resulting in isoquinoline-*N*-oxide (Sutherland et al. 1994b). In rat and rabbit liver chromosomes, the same product has been found after treatment with isoquinoline (Cowan et al. 1978; LaVoie et al. 1983).

C. Benzoquinolines

The metabolic routes of acridine and phenanthridine proceed via the epoxide/ketone intermediates acridone and 6(5*H*)-phenanthridinone (De Voogt et al. 1999), respectively, in both invertebrate and vertebrate metabolism, often resulting in dihydrodiols, hydroxides, and N-oxides (Adams et al. 1983; Kandaswami et al. 1987; LaVoie et al. 1985). While the latter three types of metabolites have been found in metabolic studies of the benzo(*f*)- and -(*h*)quinolines, the ketonic route has never been reported for these isomers. The absence of ketonic intermediates may be due to the position of the nitrogen atom, which in the (*f*)- and (*h*)-isomers may direct epoxide formation similar to what is known for methyl substituents.

Microbial degradation of acridine has been studied in river water and sediments. In river water, no degradation of acridine was observed (Cassidy et al. 1988), whereas in the laboratory under anaerobic conditions acridine was transformed by ring fission, β-oxidation, and decarboxylation into several benzyl derivatives (Knezovich et al. 1990; Fig. 4). Fungi were shown to degrade acridine into 2-hydroxyacridine and acridine *trans*-1,2-dihydrodiol via an epoxidase route (Sutherland et al. 1994a).

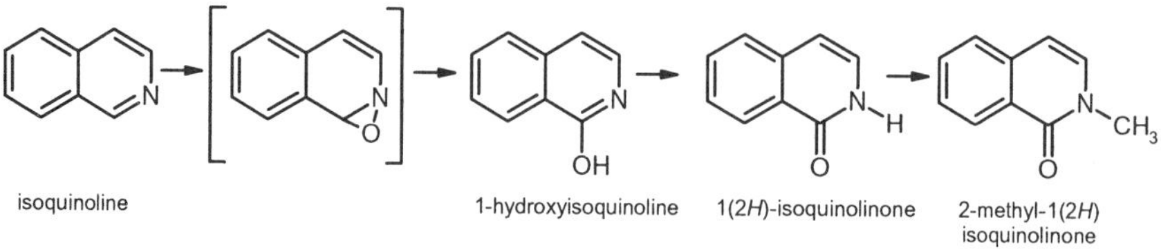

Fig. 3. Microbial degradation of isoquinoline (adapted from Pereira et al. 1987).

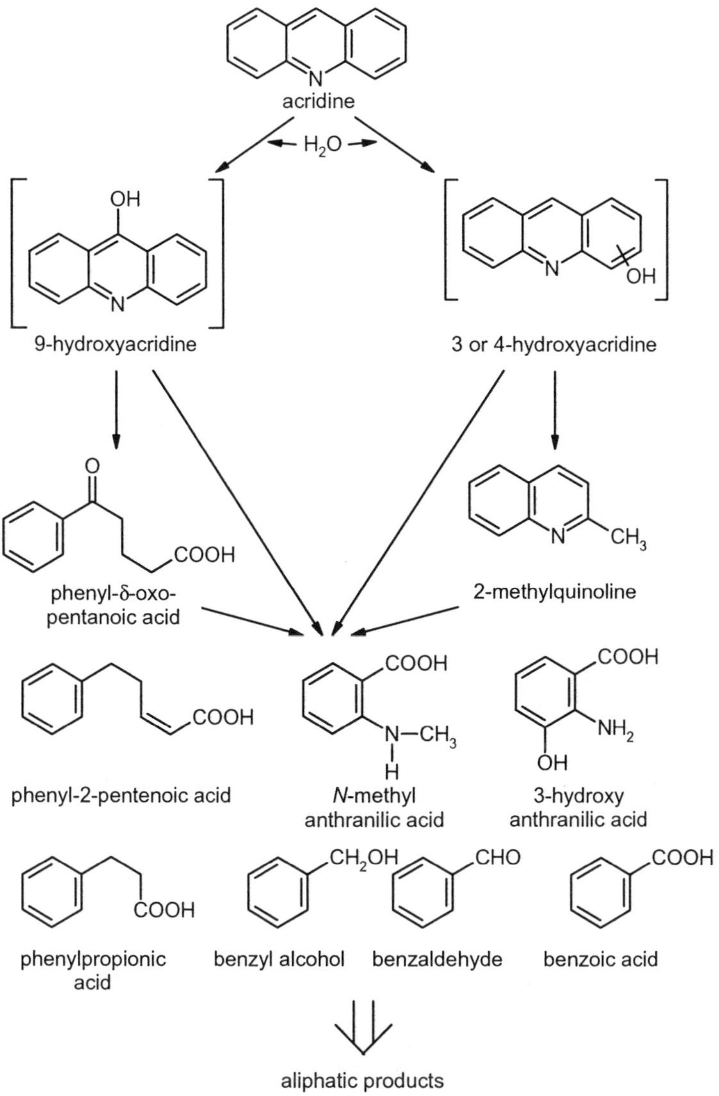

Fig. 4. Anaerobic degradation pathway of acridine (Knezovich et al. 1990).

Midges, algae, and periphyton were able to transform acridine into 9(10H)-acridone when exposed to acridine in water in the laboratory (De Voogt et al. 1999). No other metabolites could be detected. Rat liver enzymes transformed acridine *in vitro* into acridone and a nonspecified dihydrodiol compound (McMurtrey and Knight 1984). The authors suggest it to be either 2,3- or 3,4-acridine dihydrodiol.

Also for the other benzoquinolines dihydrodiols seem the most prominent transformation products (Adams et al. 1983). Polychlorinated biphenyl- (PCB)-induced rat liver homogenates transformed phenanthridine into phenanthridine-*N*-oxide, 1,2-dihydroxy-1,2-dihydrophenanthridine, and 9,10-dihydroxy-9,10-dihydrophenanthridine (Fig. 5). 6(5*H*)-Phenanthridinone and 2-hydroxyphenanthridine were identified as minor metabolites (LaVoie et al. 1985) (Fig. 5). Similar to acridine metabolism, carp were shown to transform phenanthridine into 6(5*H*)-phenanthridinone (Bleeker et al. 2001), again indicating the importance of ketone intermediates. The metabolic breakdown of benzo(*f*)- and -(*h*)quinolines by rat liver homogenate led to the identification of hydroxylated benzoquinolines and dihydrodiols (Adams et al. 1983; Kandaswami et al. 1987) (Fig. 6).

D. Azaarenes With More Than Three Condensed Aromatic Rings

For larger ring systems, microbial degradation has hardly been studied. Although dibenz(*a,h*)acridine was found to be degraded slowly by *Pseudomonas paucimobilis*, no metabolites could be identified (Siddiqi et al. 1994). No mineralization could be observed for dibenz(*a,j*)acridine (Grosser et al. 1995).

Metabolism of the larger azaarenes by vertebrates has been documented extensively because of the carcinogenicity of several compounds. The metabolism of these azaarenes closely resembles that of homoaromatic analogues, with initial formation of epoxides leading to major products such as hydroxylated and dihydrodiol compounds (Gill et al. 1986; Jacob et al. 1983; Steward et al. 1987; Wan et al. 1992; Warshawsky et al. 1985).

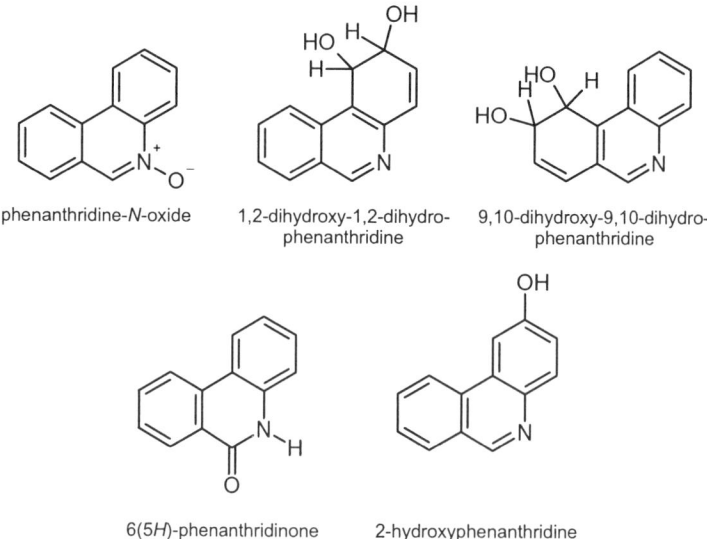

Fig. 5. Phenanthridine biotransformation products.

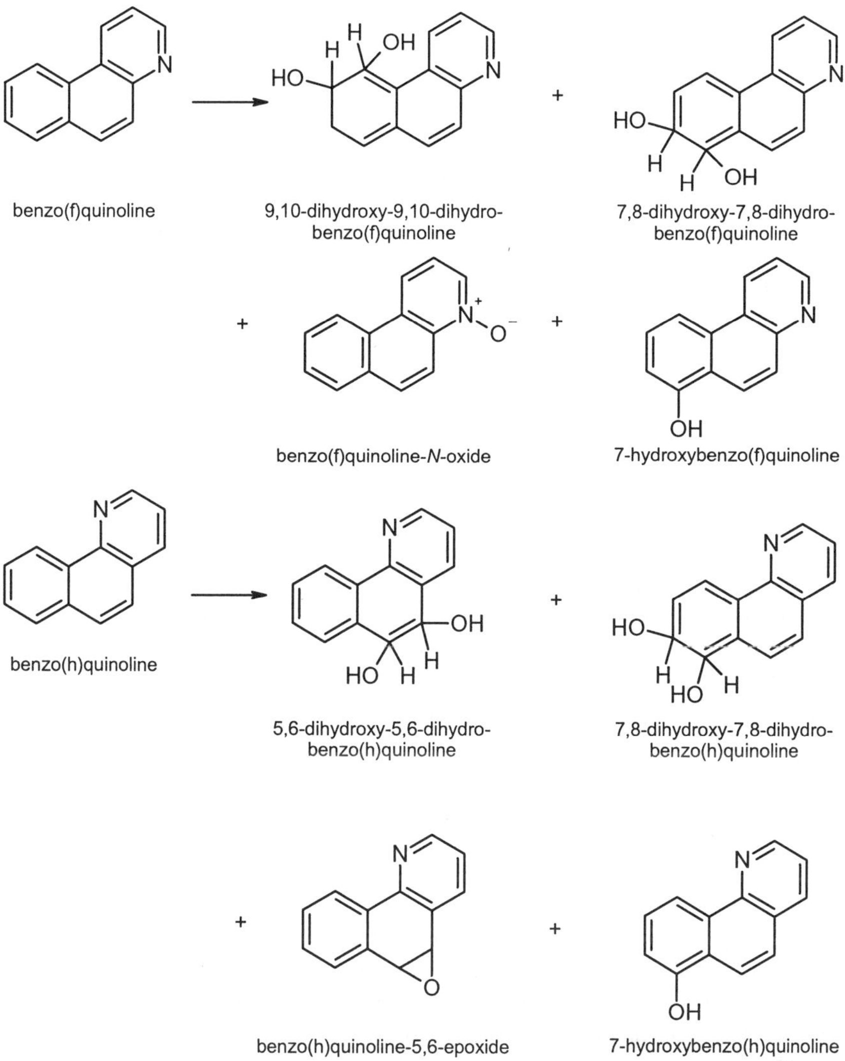

Fig. 6. Products of benzo(*f*)- and -(*h*)quinoline metabolism by rat liver homogenates.

E. Metabolites Occurring in the Environment

Determination of the degradation products of azaarenes in the environment is seldom reported. It is obvious, however, that such products must be present, as azaarenes have been reported to be present in air, water, groundwater, sediments, and suspended matter (Kozin et al. 1997; Osborne et al. 1997). Indeed, oxygenated products resulting from biotransformation of azaarenes by microorganisms have been detected in groundwater samples (Edler et al. 1997; Müller et al. 1999; Pereira et al. 1987), such as 2(1H)-quinolinone, 4(1H)-quinolinone, 1(2H)-isoquinolinone, 9(10H)-acridone, and 6(5H)-phenanthridinone.

Two- and three-ringed azaarenes can be relatively easily transformed by bacteria, fungi, invertebrates and vertebrates. The presence of the N-moiety in the smaller azaarenes leads to metabolic routes that can differ from those of the homoaromatic analogues. Major metabolic products of the azaarenes appear to be ketones and mono- or dihydroxylated azaarenes. Microorganisms can further degrade these into multiple oxygen-containing compounds or by opening up the aza-aromatic ring. Fungi and vertebrates may also produce dihydrodiol metabolites. The metabolism of the larger azaarenes in vertebrates is analogous to homoaromatic PAH because in these larger systems the N-moiety has less influence. Transformation of the larger azaarenes by microorganisms proceeds much slower if occurring at all.

III. Direct Toxicity
A. Introduction

All organic toxicants, including PAHs, induce narcosis, or baseline toxicity, to some extent. Many studies have shown that narcosis is strongly related to the lipophilicity of the compound, often expressed as the n-octanol–water partition coefficient (K_{ow}) (Chen et al. 1997; De Voogt et al. 1988; Könemann 1981; Schultz and Bearden 1998; Swartz et al. 1995). However, compound specific modes of action (i.e., other than narcosis) cause deviations from such relationships, as has clearly been demonstrated for closely related compounds such as isomers. Although they show little difference in lipophilicity, toxicity of isomers may differ by several orders of magnitude (Bleeker et al. 1999b; Kraak et al. 1997b; Kumar et al. 1989; Walton et al. 1983). Besides compound specific modes of action, deviations from structure–activity relationships may also result from photoenhanced toxicity (see Section IV).

In studying toxicity, two complementary approaches can be chosen. When species-specific sensitivities are of main interest, toxicity of one or a few compounds can be studied in many different species. Alternatively, when compound specific toxicities are assessed, one test species can be chosen for which toxicity of a whole range of compounds can be tested.

When focusing on nonsubstituted azaarenes, it becomes clear that biological test species especially are varied rather than compounds. Most research has been done in aquatic environments, although some studies report data on terrestrial life (Gissel-Nielsen and Nielsen 1996; Moir et al. 1997). Nevertheless, knowledge of the toxicity of azaarenes to aquatic species is still fragmented.

Toxicity data on azaarenes are, in about half the cases, concerned with highly substituted compounds, regarding all types of drugs and insecticides, herbicides, and fungicides. The EPA AQUIRE database (http://www.epa.gov/ecotox) shows, when searched on azaarenes and their derivatives, that most of the data (54%) are on mortality (mostly fish), and also 11% are on population parameters, with 82% considering algae, mostly growth. Other effects are rarely studied.

B. Toxicity of Quinoline and Acridine

Most research has been focused on acridine and quinoline, and only a relatively small number of species has been tested. An overall survey of effect concentrations for the various groups of aquatic species could be given only for acridine and quinoline (Fig. 7). For this survey, EC_{50} and LC_{50} values from the EPA AQUIRE database were used. This database contains 61 EC/LC_{50} values for quinoline and 69 for acridine. From Fig. 7, it is clear that effect concentrations fall in the same order of magnitude for all species groups, except for protozoa for which the quinoline toxicity tends to be less, although in this case only two studies were found, which differed greatly in effect concentration. Acridine effect concentrations are generally about 10–100 times lower than those for quinoline, most likely because of lipophilicity differences (Könemann 1981) and photoenhanced toxicity (see Section IV). These similarities in effect concentrations for all species groups could be a result of the presence of both highly sensitive and highly tolerant species.

When the data are examined at a lower taxonomic level, however, it becomes clear that this theory does not hold. For arthropods, for example, species differences in effect concentrations are still relatively small (Fig. 8), while for algae, which show a large range of effect concentrations, exposure duration (4–96 hr) appears to be more important than species differences. This result suggests that

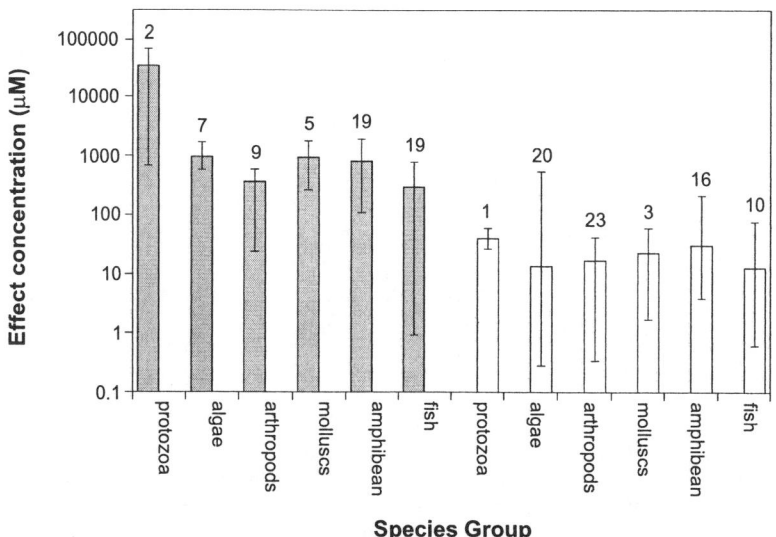

Fig. 7. Effect concentrations (both LC_{50} and EC_{50} values, in μM) of quinoline (shaded bars) and acridine (open bars) for different groups of species. Error bars indicate minimum and maximum values (when confidence limits were given the lowest or highest confidence limit is indicated). The numbers above the bars indicate the number of values used. All data are extracted from the AQUIRE database (http://www.epa.gov/ecotox).

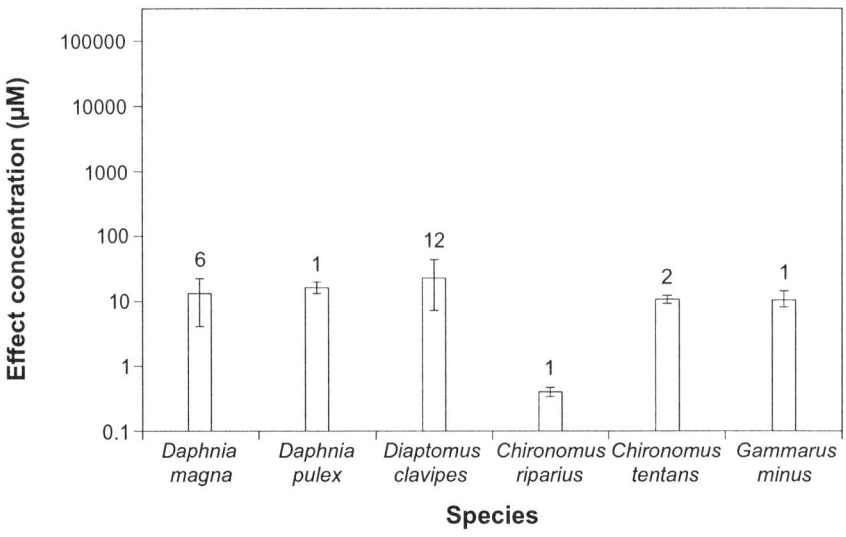

Fig. 8. Effect concentrations (both LC_{50} and EC_{50} values, in μM) of acridine for different species of arthropods. Error bars indicate minimum and maximum values (when confidence limits were given the lowest or highest confidence limit is indicated). The numbers above the bars indicate the number of values used. All data are extracted from the AQUIRE database (http://www.epa.gov/ecotox).

the underlying mechanism of azaarene toxicity is basically the same for all species considered.

C. Toxicity of Ranges of Azaarenes

Although the general approach is focused on determining the toxicity of one or a few compounds for a range of different biological taxa, there are some studies in which one biological species is exposed to a range of different azaarenes. To enable a comparison of the different compounds within and between species, species were selected for which the same endpoint was examined for at least

---

Fig. 9. Toxic effects of a range of azaarenes to different species. For each species, the same effect was scored, but between species differences in effect studied are present. In most cases LC_{50} values are given, albeit exposure durations differ between species: 96hr for *C. riparius* (Bleeker et al. 1998) and *P. reticulata* (Wegener et al. 1986), 48hr for *D. magna* (Millemann et al. 1984; Parkhurst et al. 1981; Wegener et al. 1986), and 24hr for *D. pulex* (Southworth et al. 1978). For *Scenedesmus acuminatus* the EC_{50} was based on chlorophyll a content per cell after 96hr of exposure (Van Vlaardingen et al. 1996), for *Phaeodactylum tricornutum* the EC_{50} for photosynthetic activity over 4hr was determined (Wiegman et al. 1999), and for *Dreissena polymorpha* the EC_{50} for filtration rate after 48hr was scored (Kraak et al. 1997b).

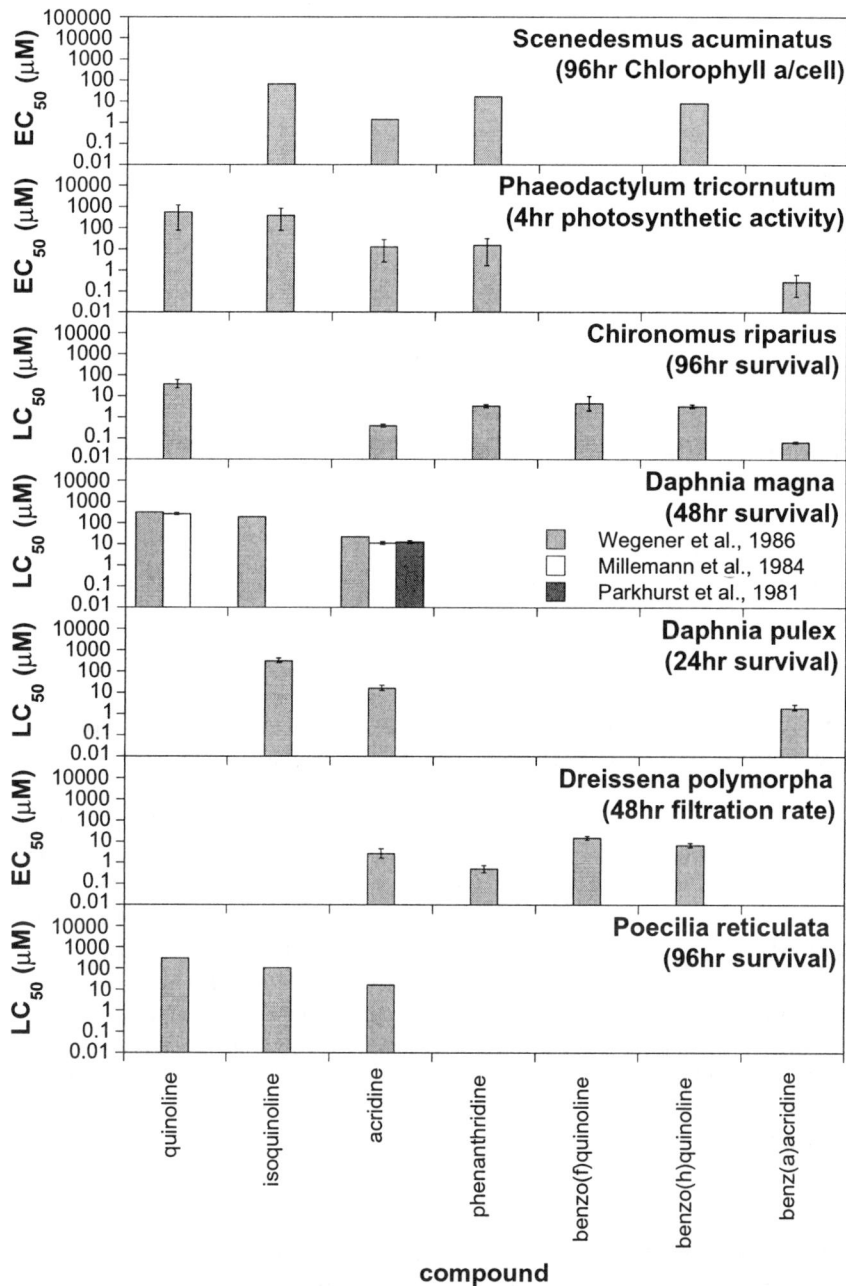

three different azaarenes. Only 7 of the 84 species mentioned in the AQUIRE database met these conditions (Fig. 9). For none of these 7 species is a complete set of all nonsubstituted 2-, 3-, and 4-ringed azaarenes available; only for *Phaeodactylum tricornutum*, *Chironomus riparius*, and *Daphnia pulex* is a series available that contains at least one of each. In these 3 species, it is shown that toxicity increases with increasing number of rings. This general pattern, however, does not explain the differences occurring between isomer structures. Even when a more detailed parameter is used (e.g., log K_{ow}), it is clear that toxicity of azaarenes cannot be explained by molecular size alone (r^2 values for the different species range from 0.007 to 0.98).

Differences in toxicity between isomers, especially structures that hardly differ in molecular or physicochemical parameters, complicate a detailed description of azaarene toxicity. Even when more specific effect parameters are used (e.g., filtration rate in mussels) large differences between isomer toxicity can occur. For acridine, one of these isomers, several studies have indicated the phototoxic susceptibility of this compound in explaining the stronger toxicity of this compound (see Section IV). In experiments with zebra mussels, however, phenanthridine is the most toxic of the three-ringed azaarenes (Kraak et al. 1997b), indicating that other molecular properties also play a role in azaarene toxicity.

D. Biotransformation and Toxicity

Oxidation products from phenanthridine and acridine, 6(5*H*)-phenanthridinone and 9(10*H*)-acridone, respectively, did not induce lethal effects in midge larvae at maximum water solubility (Bleeker et al. 1999b), concentrations similar to or greater than the acute LC_{50} of the parent compounds. Similarly, the toxicity of acridine to zebra mussels is reduced when acridine is transformed to acridone (Kraak et al. 1997a). Also, cytotoxicity in Chinese hamster ovary cells showed similar results, with the metabolite phenanthridone being less toxic than phenanthridine itself (Benson et al. 1983). Phenanthridone has also been found to reduce effects of DNA-damaging agents (Bernges and Zeller 1996; Rappeneau et al. 2000; Richardson et al. 1999). In general, oxidation to keto products appears to reduce toxicity. However, chronic exposure to low concentrations of either 6(5*H*)-phenanthridinone or 9(10*H*)-acridone led to a significant delay in emergence in *Chironomus riparius* (Bleeker et al. 1999a), suggesting other types of toxicity were being induced at lower concentrations.

E. Chronic Toxicity

For organisms with a short life cycle, it is generally agreed that a chronic toxicity test should include at least one life cycle, but other definitions are limited to one reproductive cycle or even just a 'substantial' part of either of these cycles. For this review, data were considered as chronic toxicity when test duration was at least 2 wk.

In general, chronic toxicity data for azaarenes are limited to acridine, although some chronic toxicity of quinoline to daphnids has been reported (Kühn et al. 1989) as well as some toxicity to algae for benzo(*f*)quinoline (Palmer and Maloney 1955). For derivatives such as 2,6-dimethylquinoline, some chronic data are also available (Adema et al. 1981; Hermens et al. 1984). Chronic effect concentrations for acridine vary between test species but are generally quite low (Fig. 10) (Blaylock et al. 1985; Kraak et al. 1997a; Parkhurst et al. 1981).

Attempts have been made to predict chronic toxicity from acute test results by deriving acute-to-chronic ratios (ACRs) (Kenaga 1982; Länge et al. 1998; Mayer 1990; Roex et al. 2000). Such predictions inherently assume a similar kind of action during both acute and chronic exposure. Hence, deviations from these predictions are usually explained by other, more specific effects that are expressed at lower exposure concentrations during a longer exposure time, yet ACRs are used in risk assessment to come up with standard safety factors. Even from a very limited set of chronic data on azaarenes, however, it is clear that the often-used safety factor of 10 (Van Leeuwen et al. 1992) provides protection only for a limited set of species (Fig. 10). For quinoline, even a safety factor of 100 appears to be insufficient, and for *Chironomus riparius* the same holds for acridine toxicity (Bleeker et al. 1999a). These results show that even when ACRs are calculated only for compounds with similar modes of action (Roex et al. 2000), their use in risk assessment should be applied with great care.

IV. Photochemical Transformation and Phototoxic Effects of Azaarenes
A. Introduction

Many PAHs present in aqueous media can undergo photochemical transformation under the influence of radiation. In the natural environment, the source of radiation is sunlight, which has a broad spectral distribution in the range of 290 to 800 nm. Several PAHs absorb sunlight in the visible region (>400 nm) but absorb ultraviolet radiation (280–400 nm) to a greater extent; the portion of this spectrum entering aquatic media causes a diverse number of chemical reactions in natural and xenobiotic compounds (Arfsten et al. 1996). The rate of transformation by light depends on the overlap between the UV absorption spectrum of the compound (ε_λ) and the incident solar light intensity at wavelength λ (W_λ) (Leifer 1988).

B. Mechanisms and Kinetics of Photochemical Reactions of (N)PAHs

UV radiation can alter the structures of PAHs by essentially two routes (Fig. 11): photosensitization reactions initiated by the PAHs and photomodification of the PAHs (Larson and Berenbaum 1988). Photomodification structurally alters PAHs to a variety of oxygenation products via unstable endoperoxide or peroxide intermediates (David and Boule 1993; Huang et al. 1993, 1995; Zepp and Schlotzhauer 1979).

In photosensitization reactions, the influence of light on the fate of PAHs in

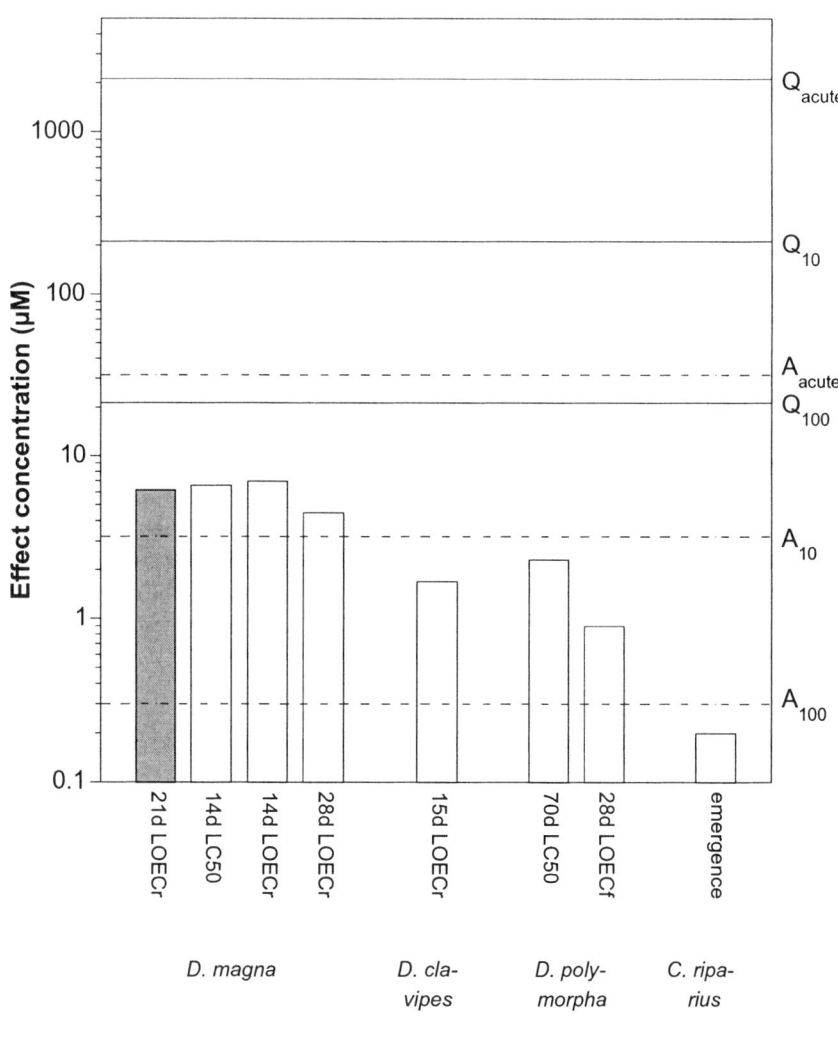

Fig. 10. Comparing acute and chronic toxicity. Most chronic effects are expressed either as lowest observed effect concentrations for reproduction (LOECr), or for filtration rate (LOECf) or as LC_{50} values. Emergence: significant delay in average day of emergence. Test durations differ for the different effects (14–70d). *D. magna: Daphnia magna; D. clavipes: Diaptomus clavipes; D. polymorpha: Dreissena polymorpha; C. riparius: Chironomus riparius*; Q_{acute}: average acute LC_{50} for aquatic species exposed to quinoline (derived from the AQUIRE database); A_{acute}: average acute LC_{50} for aquatic species exposed to acridine (derived from the AQUIRE database); Q_{10}: safety factor of 10 for quinoline; Q_{100}: safety factor of 100 for quinoline; A_{10}: safety factor of 10 for acridine; A_{100}: safety factor of 100 for acridine (see text for further details).

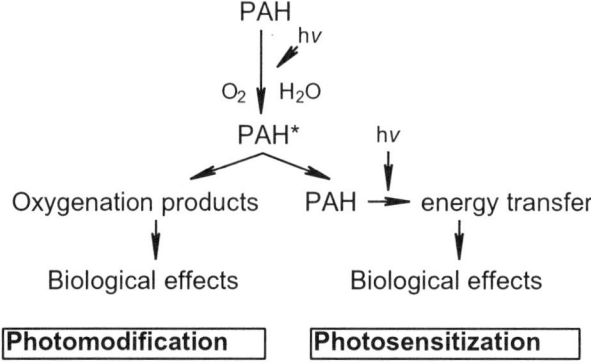

Fig. 11. Schematic view of PAH chemical reaction in the water column. Modified after Foote (1991).

water is to a large extent determined by the presence or absence of reactive species (Larson and Berenbaum 1988). In the presence of oxygen, PAH–oxygen complexes will be formed, also called charge transfer or CT complexes, which can be excited by photons. On excitation, the CT complex can dissociate along several pathways (Kochany and Maguire 1994; Onodera et al. 1985):

$$^1PAH \text{ (singlet ground state)} + h\nu \rightarrow {}^1PAH^* \text{ (singlet excited state)}$$
$$^1PAH^* \rightarrow {}^3PAH^* \text{ (triplet excited state)}$$
$$^3PAH^* + {}^3O_2 \rightarrow {}^1PAH + {}^1O_2$$
$$^1PAH + {}^1O_2 \rightarrow \text{oxygenation products}$$

The excited singlet-state PAH can undergo intersystem crossing to the excited triplet state, after which it can react with ground triplet-state oxygen ($^3O_2$) to form $^1O_2$ (Larson and Berenbaum 1988; Onodera et al. 1985). In contrast to the lifetimes of singlet-state molecules, triplet-state lifetimes of PAHs are sufficiently long for diffusion-limited reactions with $^3O_2$ and are therefore effective photosensitizers (Krylov et al. 1997). It should be noted however, that triplet-state PAH can excite ground state oxygen (22 kcal/mol required) only if the triplet–singlet conversion of the PAH leads to an energy gain (>38 kcal/mol). Excited oxygen, although having an extremely short life span, will attack those carbon atoms in the benzene ring opposite (i.e., para oriented) to those having the highest electron density of the HOMO (highest occupied molecular orbital) resulting in 1,4-adducts, also called endoperoxides (Yamaguchi et al. 1985). Endo-compounds, on their turn, can dissociate into the parent PAH and $^1O_2$. Formed within organisms, $^1O_2$ is capable of oxygenating or oxidizing many different biomolecules, altering their chemical structure, and consequently inhibiting or inactivating them (Larson and Berenbaum 1988).

The kinetics and products of photochemical reactions of (N)PAHs in aquatic environments have been reviewed by Kochany and Maguire (1994). Compared to homocyclic PAHs, little work has been done on photochemical oxidation of azaarenes in water. Nevertheless, photochemical reactions are thought to be more important for azaarenes than for PAHs because solubilities of azaarenes are 1,000 to 10,000 times higher than those for PAHs (Kochany and Maguire 1994).

The exposure of aromatic compounds to UV radiation depends on the depth and the composition of the water (Leifer 1988). In water, UV radiation is absorbed and scattered by dissolved compounds, pigments, and suspended material, especially humic acids, which form a part of the dissolved organic matter. Thus, high concentrations of organic matter lead to a strong decrease of UV with increasing depth and consequently to a decrease of photochemical reactions of PAHs. Increasing rates of photodegradation of azaarenes were also found in the presence of humic acids, suggesting that humic acids can act as sensitizers (Kochany and Maguire 1994).

Besides light attenuation, photolysis rates are determined by turbidity, presence of sensitizers, and for azaarenes especially by the pH of the aquatic environment. In the aromatic system of azaarenes, the electron density is higher than for PAHs because of the unbonded electron pair on the nitrogen atom. In acidified water, the nitrogen atom will be protonated (at pH 4.4, 70% of quinoline is protonated; Kochany and Maguire 1994), lowering the electron density within the molecule and thereby decreasing photolysis rates (Kochany and Maguire 1994; Mill et al. 1981).

In laboratory studies, azaarenes dissolved in pure water degraded rapidly under short-wave radiation, UV-B radiation being more effective than UV-A (Kochany and Maguire 1994; Mill et al. 1981; Wiegman et al. 1999). In outdoor experiments, however, the photochemical reactions proceeded more slowly (Mill et al. 1981; Wiegman et al. 1999), as is shown in Fig. 12 in which photolysis rates of acridine under different light sources are compared. Acridine degraded faster when exposed to laboratory UV radiation than to sunlight (52°21' N, The Netherlands, May 23–27, 1997). Compared to sunlight with a broad spectral composition, laboratory UV radiation contains a small range of wavelengths with relatively high intensities. Besides overlap between absorption spectrum of the compound ($\Sigma I \epsilon \lambda$) and the incident light, the photolysis rates of compounds are proportional to the efficiency of direct photochemical reaction or to quantum yield (ϕ). For acridine and quinoline, the quantum yields—the ratio between the number of molecules undergoing photochemical reaction and the number of photons absorbed—was lower in photolysis experiments conducted under natural light than under lamps with a smaller region of UV (Mill et al. 1981; Wiegman et al. 1999). Although for PAHs it is assumed that quantum yields are independent of wavelength (Leifer 1988; Zepp 1978), the opposite is found for azaarenes, ranging from two-ringed to five-ringed structures, and the homocyclic PAH fluoranthene (Leifer 1988; Wiegman et al. 1999; Zepp 1978). The mechanistic pathway for photodegradation of azaarenes is still partially unclear. In contrast to the photoreactions of some PAHs (Leifer 1988; Zepp 1978), for azaarenes different excitation processes seem to play a role (Wiegman et al. 1999).

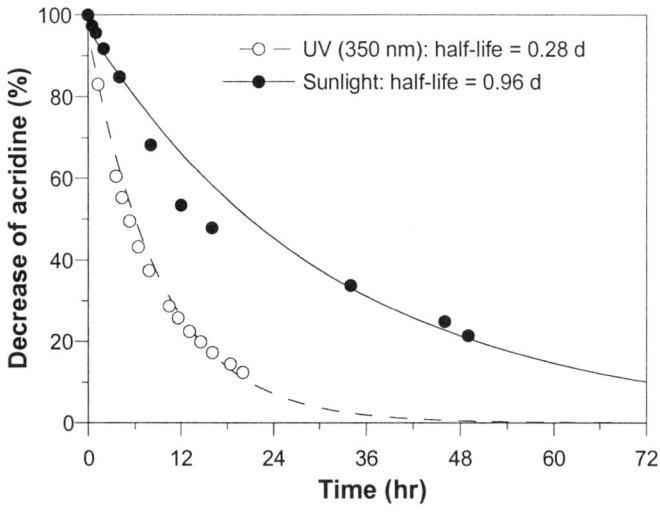

Fig. 12. Decrease of acridine concentration (%) in milliQ-water by UV ($C_0 = 1.8$ mg/L, 0.033% v/v solvent DMSO) and by sunlight (1500 FE, $C_0 = 1.8$ mg/L, 1% v/v solvent acetonitrile) plotted against time (hr).

Compared to PAHs, e.g., anthracene, phenanthrene, and benzo(*a*)pyrene (Mallakin et al. 1999; McConkey et al. 1997; Mill et al. 1981), little attempt has been made to identify the whole series of photodegradation products of azaarenes. Photoproducts determined for indoles were *O*-acylaminoketones (Picel et al. 1987). For quinoline some photoproducts were tentatively determined, mainly hydroxyquinolines. For acridine, the main photoproduct found was 9(10*H*)-acridone (De Voogt et al. 1999; Wiegman et al. 1999). For quinoline and acridine, these products are the same as those formed in biological transformation involving monooxygenases (De Voogt et al. 1999; Kaiser et al. 1996; see Section II).

C. Phototoxic Effects of (N)PAHs and Their Photoproducts on Aquatic Organisms

The activation by light of PAHs into reactive and toxic products is defined as photoenhanced toxicity. Kagan et al. (1987) noted that the degree of photoenhanced toxicity of PAHs under laboratory UV radiation is related to UV absorption characteristics for each compound. The absorption peak of the phototoxic PAH is found to be in the UV-A range of light (320–400 nm) with a high amount of energy, and of the nonphototoxic PAHs near the UV-B range of light (<320 nm), with a relatively small amount of energy. Holst and Giesy (1989) and Ankley et al. (1995) observed that increasing UV intensities coincide with increasing photoenhanced toxicity of PAHs to aquatic organisms. The energy difference between the highest occupied molecular orbital and the lowest unoccupied molecular orbital (HOMO–LUMO gap) of PAH molecules accurately explained the persistence, light absorption, and photoenhanced toxicity of PAHs. The HOMO–LUMO gap of PAHs defines the energy that is necessary to elevate an electron from the HOMO to the LUMO state. The PAHs with larger HOMO–LUMO gaps absorb smaller wavelengths (UV) of light (greater energy) and are more reactive; the PAHs with small HOMO–LUMO gaps absorb less energy (more visible range of light) and are photochemically less reactive (Mekenyan et al. 1994a; Veith et al. 1995). Photoenhanced toxicity occurs for PAHs with ground-state HOMO–LUMO gap energies in the range of 6.5 to 8.0 eV with maximum effect around 7.2 ± 0.4 eV (Mekenyan et al. 1994a,b). PAHs falling within this highly 'phototoxic window' are anthracene, pyrene, benzo(*a*)anthracene, benzo(*e*)pyrene, benzo(*a*)pyrene, dibenz(*a,h*)anthracene, and perylene (Table 1) (Boese et al. 1998; Mekenyan et al. 1994a; Veith et al. 1995).

Under standard laboratory light conditions, most of the four- and five-ringed PAHs are not acutely toxic at or below their aqueous solubilities. However, in the presence of solar radiation, a number of 'phototoxic' PAHs have been found to be acutely toxic to aquatic organisms at concentrations well below their solubilities (Gala and Giesy 1993, 1994; Holst and Giesy 1989; Kagan et al. 1985; Pelletier et al. 1997; Wernersson and Dave 1997).

Most research concerning photoenhanced toxicity in the aquatic environment has been focused on acute toxicity of PAHs to freshwater species (reviewed by Arfsten et al. 1996), e.g., the water flea *Daphnia magna* (Kagan et al. 1985; Wernersson and Dave 1997), the mosquito *Aedes aegypti* (Kagan et al. 1985), the midge *Chironomus tentans* (Ankley et al. 1994), the oligochaete *Lumbriculus variegatus* (Ankley et al. 1995), the fish species *Pimephales promelas* (Kagan et al. 1987) and *Lepomis macrochirus* (Oris and Giesy 1985, 1986), the alga *Selenastrum capricornutum* (Gala and Giesy 1993, 1994), and the macrophyte *Lemna gibba* (Huang et al. 1993). Phototoxic responses of PAHs vary between test species and compounds. In algae, highly colored pigments can protect against phototoxic effects of PAHs (Gala and Giesy 1993). Because most algae and plants contain these pigments, they generally show a higher resistance to phototoxic PAHs than do invertebrates and fish (Arfsten et al. 1996).

Besides acute effects of phototoxic PAHs, the chronic effects of anthracene under UV-A radiation on *Daphnia magna* have been reported by Holst and Giesy (1989). Exposure to anthracene in the presence of ecologically relevant UV-A intensities decreased the survival and fecundity of *D. magna* at concentrations well below their aqueous solubility limits, indicating that low PAH concentrations can adversely affect population dynamics in the environment.

A few experiments with field-collected sediments containing a combination of different PAHs have been conducted (Arfsten et al. 1996). The benthic organisms *Hyalella azteca* and *Lumbriculus variegatus* in particular accumulated PAHs from the test sediments, and subsequently these PAHs were photoactivated by UV in both laboratory and field experiments (Ankley et al. 1994; Monson et al. 1995).

Table 1. HOMO-LUMO gap energies and log K_{ow} of several PAHs and azaarenes (*italic*). (N)PAHs with HOMO-LUMO energy values near 7.2 ± 0.4 eV are phototoxic (**bold**).

Compound	Rings	Log K_{ow}	HOMO-LUMO
Naphthalene	2	3.32[c]	10.06[h]/10.16[b]
Quinoline	2	*2.03*[i]	*8.72*[g]
Isoquinoline	2	*2.08*[i]	*8.46*[j]
Fluorene	3	4.32[e]	8.49[b]/8.51[d]
Acenaphthene	3	3.9[h]	8.24[h]
Anthracene	3	**4.57**[c]	**7.28**[b,d]
Phenanthrene	3	4.67[c]	8.20[h]/8.21[b]
Acridine	*3*	***3.20/3.41***[c]	***7.53***[b,g]
Benzo(f)quinoline	*3*	*3.43*[g]	*8.28*[g]
Benzo(h)quinoline	*3*	*3.43*[g]	*8.18*[g]
Phenanthridine	*3*	*3.48*[g]	*8.37*[g]
Fluoranthene	4	5.22[c]	7.68[h]/7.70[b]
Pyrene	4	**4.9**[h]	**7.24**[b,h]
Chrysene	4	5.79[c]	7.69[b]
Benzo(*a*)anthracene	4	**5.90**[h]	**7.39**[b]
Benzo(*b*)anthracene	4	5.90[i]	6.52[b]
Benz(a)acridine	*4*	*4.49*[g]	*7.63*[g]
Benz(c)acridine	*4*	*4.49*[g]	*7.56*[g]
Benzo(*b*)fluoranthene	5	**5.8**[h]	**7.59**[h]
Dibenzo(*a,h*)anthracene	5	**6.40**[e]	**7.45**[b]
Benzo(*e*)pyrene	5	**6.44**[f]	**7.36**[b]
Benzo(*a*)pyrene	5	6.10[c]	6.81[b]
Perylene	5	6.44[f]	6.70[b]/6.71[d]
Dibenz(a,i)acridine	*5*	*5.67*[k]	*6.81*[k]
Dibenz(c,h)acridine	*5*	*6.27*[e]	*7.64*[k]

[a]Neff (1985), [b]Mekenyan et al. (1994a), [c]Kalf et al. (1995), [d]Veith et al. (1995), [e]Helweg et al. (1997), [f]Baumard et al. (1998), [g]Bleeker et al. (1998), [h]Boese et al. (1998), [i]Hansch et al. (1995), [j]Wiegman et al. (2001), [k]calculated with Hyper Chem (Hypercube Inc, Version 6.1).

Photoenhanced toxicity of azaarenes in aqueous environments has been poorly investigated. Acridine, benz(*a*)acridine, and benz(*c*)acridine are the only azaarenes that can display photoenhanced toxicity in aqueous surroundings (see Table 1). Of these compounds, acridine will be the most pronounced phototoxic compound because the HOMO-LUMO gap of acridine falls in the highly phototoxic region of 7.2 ± 0.4 eV (Mekenyan et al. 1994a). Phototoxicity of acridine is indeed reported for the freshwater ciliate *Tetrahymena pyriformis*, the water flea *Daphnia magna*, the fathead minnow *Pimephales promelas*, and several algal species (Table 2) (Bleeker et al. 1998; Dijkman et al. 1997; Newsted and Giesy 1987; Oris and Giesy 1987; Sinks et al. 1997; Wernersson and Dave 1997; Wiegman et al. 2001).

Table 2. An overview of the actual effect concentrations of acridine (mg/L) for different freshwater and marine species (M), obtained from laboratory tests under fluorescent light, mercury light or irradiated with UV.

Effects of acridine to	mg/L	Light[a]	Endpoints	References
Fish				
Pimephales promelas	2.24–2.90	no UV (dark)	96hr, lethality	1,3
	0.53	UV-B, UV-A	4.3hr, LT_{50}	2
Invertebrates				
Chironomus tentans	1.86, 1.96	no UV (dark)	48hr, lethality	1,4
Chironomus riparius	0.07	mercury	96hr, lethality	5
Daphnia pulex	1.71–2.92	no UV	24hr, lethality	6
Daphnia magna	1.18–3.10	no UV	48hr–14d, lethality	1,3,7
	0.168	2h UV-A	24hr, immobility	8
	0.44	UV-B, UV-A	0.9hr, LT_{50}	9
Diaptomus clavipes	1.55–1.59	no UV	17hr–85hr, lethality	10
Dreissena polymorpha	0.50–0.96	mercury	48hr, filtration	11,12
Gammarus minus	1.87	no UV	48hr, lethality	1
Physa gyrina	11	no UV	48hr, lethality	1
Tetrahymena pyriformis	7.35	no UV	72hr, growth	13
Algae				
Chlamydomonas eugametos	0.78–0.84	mercury	72hr, growth	14
Dunaliella tertiolecta (M)	0.38–0.42	mercury	72hr, growth	15
Navicula salinarum (M)	0.33–0.82	mercury	96hr, growth	14
Nitzschia sigma (M)	0.08–0.13	mercury	96hr, growth	14
Nitzschia palea	20.8	no UV	4hr, photosynthesis efficiency	1
Scenedesmus acuminatus	0.26–0.44	mercury	72hr–96hr, chlorophyll-a, growth	14,16
Selenastrum capricornutum	0.27–0.78	mercury	72hr, growth	14
	4.7–20.0	no UV	4hr, photosynthesis efficiency	17,1
Staurastrum chaetoceras	0.22	mercury	96hr, growth	14
Staurastrum manfeldtii	0.35–0.46	mercury	96hr, growth	14
Phaeodactylum tricornutum (M)	2.26	no UV	4hr, PSE	18

[a]Mercury light contains a small portion of UV radiation.

References: 1, Millemann et al. (1984); 2, Oris and Giesy Jr. (1987); 3, Blaylock et al. (1985); 4, Cushman and McKamey (1981); 5, Bleeker et al. (1998); 6, Southworth et al. (1978); 7, Parkhurst et al. (1981); 8, Wernersson and Dave (1997); 9, Newsted and Giesy (1987); 10, Cooney and Gehrs (1984); 11, Kraak et al. (1997a); 12, Kraak et al. (1997b); 13, Schultz et al. (1980); 14, Dijkman et al. (1997); 15, Wiegman et al. (2000); 16, Van Vlaardingen et al. (1996); 17, Giddings (1979); 18, Wiegman et al. (1999).

In general, as for homocyclic PAHs (Kagan et al. 1987), the degree of phototoxicity is related to the UV absorption characteristics of the azaarenes. Irradiated with light in the spectral region of 300–410 nm, the absorption peak of acridine is around 355 nm (Wiegman et al. 1999). For both benzacridines, the absorption peaks are between 360 and 390 nm (Wiegman et al. 1999), indicating that the UV-A region of light (320–400 nm) especially is responsible for photoenhanced toxicity of azaarenes, as is demonstrated for acridine to *Daphnia* (Wernersson and Dave 1997) and to *Tetrahymena* (Sinks et al. 1997). UV-A constitutes a much larger fraction of sunlight at the earth's surface and in the water column than the more reactive UV-B, which indicates that, as UV-B is hardly present in the water column, UV-A is the most important light source causing photoenhanced toxicity of PAHs and azaarenes in aquatic environments.

Photosensitization of PAHs is restricted to PAHs exhibiting a HOMO–LUMO gap window of 7.2 ± 0.4 eV only whereas photomodification of PAHs is not. Also, PAHs with HOMO–LUMO energies differences outside this HOMO–LUMO gap window can be modified to more toxic photoproducts (Huang et al. 1993; Ren et al. 1994). Although most studies demonstrated that electronic forces are responsible for phototoxicity of (N)PAHs to aquatic organisms (Ankley et al. 1997; Gala and Giesy 1994; Mekenyan et al. 1994a; Veith et al. 1995), the photomodification of PAHs into oxygenated products has been proven to increase the environmental hazard of PAHs as well (Huang et al. 1993, 1995, 1997a,b; Mallakin et al. 1999; McConkey et al. 1997; Ren et al. 1994). Huang et al. (1997a) observed that some photomodified PAHs have a unique mechanism of toxicity to photosynthesis of plants. Especially, the oxygenation products seemed to block the electron transport between photosystem I and II, inhibiting photosynthesis of the duckweed *Lemna gibba* (Huang et al. 1997a) and natural phytoplankton assemblages (Marwood et al. 1999).

The effect of eight different azaarenes and their phototransformation products on the photosynthesis of the marine diatom *Phaeodactylum tricornutum* was determined by Wiegman et al. (1999). Clear dose–response relationships were only established for unmodified two- and three-ringed structures and for benz-(*a*)acridine. The effect concentrations of benz(*c*)acridine and two dibenzacridines were above water solubilities. Next, azaarene solutions were irradiated with, respectively, UV-B (250–350 nm) and UV-A (305–410 nm) lamps, until at least 75% of initial parent azaarene concentration had been modified before the toxicity tests. These solutions were tested in a concentration series similar to the unmodified azaarenes. During the test, fluorescence light (>400 nm) was used to avoid photosensitization. The toxicity of azaarene solutions modified with UV-B was lower than or equaled the toxicity of unmodified azaarenes. Thus, toxicity of UV-B modified azaarenes was primarily caused by the parent azaarenes. UV-A radiation increased toxicity of quinoline and isoquinoline solutions (two-ringed azaarenes) almost two orders of magnitude but that of acridine and phenanthridine solutions only modestly. This result clearly demonstrates that toxicity of azaarenes that are not susceptible to photoactivation can also be enhanced by UV, which stresses the importance of photomodification as a pathway of phototoxicity.

V. Genotoxicity and Carcinogenicity
A. Introduction

A mutagen is a chemical that causes heritable changes in the genome. Mutagens may act *directly* by changing the nucleotide sequence of DNA (e.g., frameshifts, point mutations) or by inducing chromosome aberrations (e.g., aneuploidy); this is referred to as immediate genotoxicity. Mutagens may also act *indirectly*, whereby the biotransformation product of the compound is responsible for the DNA damage; this is mediated genotoxicity (Appel et al. 1990; Becker et al. 1996). In the case of indirect mutagenicity, enzymatic detoxification pathways lead to the metabolic activation of otherwise nonthreatening amounts of xenobiotics.

In a strict sense, carcinogenesis signifies the development of an epithelial tumor only, a carcinoma (Appel et al. 1990). A compound qualifies as a carcinogen if it leads to either increased incidence or earlier development of tumors, development of different tumor types, or increased multiplicity of tumors, compared with control (Klaassen et al. 1996). There is, however, controversy over whether only malignant tumors should be accepted for the definition, or if increases in benign tumor frequency also constitute a carcinogenic effect. Appel et al. (1990) have defined chemical carcinogenesis as "a process in which... uncontrolled and irregular *de novo* formation of tissue (neoplasia) takes place." They suggest a standardized terminology whereby cancerogenesis would refer to induction of malignant tumors and tumorigenesis (or oncogenesis) to the induction of either benign or malignant tumors.

In contrast to various biological carcinogens, chemical induction of cancer always follows a direct or indirect mutagenic event (Becker et al. 1996). Therefore, tests to screen for mutagens are often used to infer information on the carcinogenicity of the compounds as well. In the widely used Ames test, 90% of compounds that test positive for mutagenicity also induce cancer when tested in animals (Becker et al. 1996). However, carcinogenicity differs from mutagenicity in that it is a multistage process, with initiation and promotion phases, which often requires several mutation events all in the same cell and all escaping DNA-repair mechanisms before a tumor is induced (Appel et al. 1990).

Mutagenicity is often assayed using the widely known Ames test, or with the Mutatox^{TM} test. Other genotoxicity tests, such as the SOS-chromotest and UmuC-test, are usually applied to generate data on extracts from the field or on PAH mixtures, rather than individual compounds. The Ames test is based on a system using a histidine-dependent strain of bacteria, *Salmonella typhimurium*, which in the absence of histidine will only grow after the histidine-independence is restored by mutation (Ames et al. 1973b).

The Mutatox genotoxicity test assays the ability of potential mutagens to cause dark mutants of the luminescent bacterium, *Vibrio fisheri*, to revert to the luminescent state by measuring increase in light intensity emitted by the bacteria (Johnson 1992). In both tests, S9 rat liver fraction is added if the mutagenicity of metabolites is also being tested. Enzymes from the liver are specialized in detoxification of xenobiotics, and its addition in the test mimics the transformation of potential mutagens to give a better impression of the *in vivo* effects (Ames et al. 1973a).

Azaarenes, because of their planar fused-ring structures, are capable of intercalating between the bases of DNA. Highly electrophilic intermediates formed during biotransformation (e.g., diol epoxides) can easily bind strongly (but reversibly) with nucleophilic carboxyl, amino, and sulfhydryl groups on nucleic acids or proteins. When a covalent bond forms between a carbonium ion intermediate and the DNA helix, a DNA adduct forms, which leads to a disruption of the DNA configuration and possibly to mutagenesis and cancer.

B. Quinoline and Isoquinoline

Quinoline has proved to be mutagenic in the *Salmonella typhimurium* test (Ho et al. 1981; Santodonato and Howard 1981; Seixas et al. 1982), although only after metabolic activation, indicating that especially its metabolites are mutagenic, rather than quinoline itself. Isoquinoline, in contrast, did not show mutagenicity at all (Santodonato and Howard 1981; Seixas et al. 1982). Considering the active metabolite of quinoline, Seixas et al. (1982) provided evidence that neither quinoline-*N*-oxide nor quinoline-2,3-oxide was likely to be this metabolite. Tada et al. (1980) suggested the mutagenic mechanism to be the covalent binding to DNA bases via either the 2,3- or the 3,4-epoxy derivative of quinoline, which indicates that quinoline-3,4-oxide is the most likely mutagenic metabolite.

In contrast to these results from the Ames test, the Mutatox^{TM} test showed both quinoline and isoquinoline to be directly genotoxic (Bleeker et al. 1999b; Wiegman et al. 1998), although the least observable effective concentration (LOEC) values are quite high and isoquinoline is again less genotoxic than quinoline. The reason for these contrasting results is most likely due to the response of the Mutatox^{TM} test to blocking or alteration of the luminescence repressor.

Quinoline was shown to induce liver cancer (Hirao et al. 1976; LaVoie et al. 1988), and it was suggested that quinoline is activated in the liver only. In tests for isoquinoline carcinogenicity, results were negative (Santodonato and Howard 1981).

C. Benzoquinolines

Three-ringed structures have been reported to be nontumorigenic (Kumar et al. 1989; LaVoie et al. 1988; Sutherland et al. 1994a), but genotoxic effects have been found. Acridine differs from the other benzoquinolines in being directly mutagenic both in the *Salmonella typhimurium* test (Seixas et al. 1982) and in the Mutatox^{TM} assay (Bleeker et al. 1999b). In contrast, one of its metabolites, 9(10*H*)-acridone, is much more genotoxic in the Mutatox^{TM} test than acridine itself (Bleeker et al. 1999b), suggesting that in rat liver 9(10*H*)-acridone is not a major transformation product. Because this metabolite has been found in groundwater (Edler et al. 1997; Müller et al. 1999; Pereira et al. 1987), this indicates that genotoxic risk of acridine in the field may be higher than expected from laboratory experiments with rat liver homogenate.

The Mutatox™ also showed direct genotoxicity of the other benzoquinolines tested, phenanthridine and benzo(*h*)quinoline (Bleeker et al. 1999b). The Ames test results for benzoquinolines are less unambiguous. Although phenanthridine and benzo(*f*)quinoline were mutagenic after metabolic activation with S9 liver fraction (Adams et al. 1983; Seixas et al. 1982), the strain of *Salmonella typhimurium* used appeared to be determining whether benzo(*h*)quinoline was mutagenic or not (Adams et al. 1983; Kumar et al. 1989; compared to Seixas et al. 1982). Also, differences in preparation and condition of the S9 used may influence the outcome of these tests (De Maagd and Tonkes 2000).

Similar to homocyclic PAHs, it may be expected that diol epoxides will be the active transformation products in mutagenesis of azaarenes. For benzo(*h*)quinoline and benzo(*f*)quinoline, this indeed appears to be the case (Kandaswami et al. 1987; Kumar et al. 1989). The difference in structure, however, results in strong differences in mutagenesis of the two compounds. For benzo(*h*)quinoline, benzo(*h*)quinoline-7,8-diol and *anti*-benzo(*h*)quinoline-7,8-diol-9,10-epoxide are suggested to be the major proximate and ultimate mutagenic metabolites, respectively (Fig. 13) (Kumar et al. 1989); this shows that the bay region theory of the metabolic activation of PAHs (Jerina and Lehr 1977) can be extended to the mutagenesis of benzo(*h*)quinoline. For benzo(*f*)quinoline, however, it was shown that the parent compound is significantly more mutagenic than the dihydrodiol precursor to its bay region dihydrodiol epoxide (Kumar et al. 1989). In this case, the formation of benzo(*f*)quinoline-7,8-dihydrodiol and its conversion to the bay region diol epoxide may not be the principal mechanism by which benzo(*f*)quinoline exerts its mutagenic activity. This theory is further supported by the weak intrinsic mutagenic activity of *cis*-benzo-(*f*)quinoline-7,8-diol-9,10-epoxide and benzo(*f*)quinoline-H_4-9,10-epoxide (Kumar et al. 1989). The location of the nitrogen atom in either the K region (benzo-(*f*)quinoline) or the bay region (benzo(*h*)quinoline) appears to be the key factor in explaining these differences.

For phenanthridine, somewhat different results have been reported. Its metabolite phenanthridone has been reported as a proximate and ultimate mutagen in the Ames test (Benson et al. 1982, 1983). The direct mutagenicity of this metabolite can be enhanced by further metabolism of the compound, possibly into phenol or dihydrodiol derivatives. Other metabolites, however, may still contribute to the mutagenicity of phenanthridine. In particular, the epoxides that are precursors to the 1,2- and 9,10-dihydrodiols of phenanthridine could be associated with the mutagenic activity of phenanthridine (LaVoie et al. 1985). These dihydrodiols again confirm the bay region theory of mutagenicity of PAHs. Data from the Mutatox™ test only partly support these Ames test data. 6(5*H*)-Phenanthridinone did show direct genotoxicity in the Mutatox™ test, but to a similar extent as phenanthridine, while in the Ames test phenanthridone shows a stronger mutagenicity than phenanthridine does, both with and without S-9.

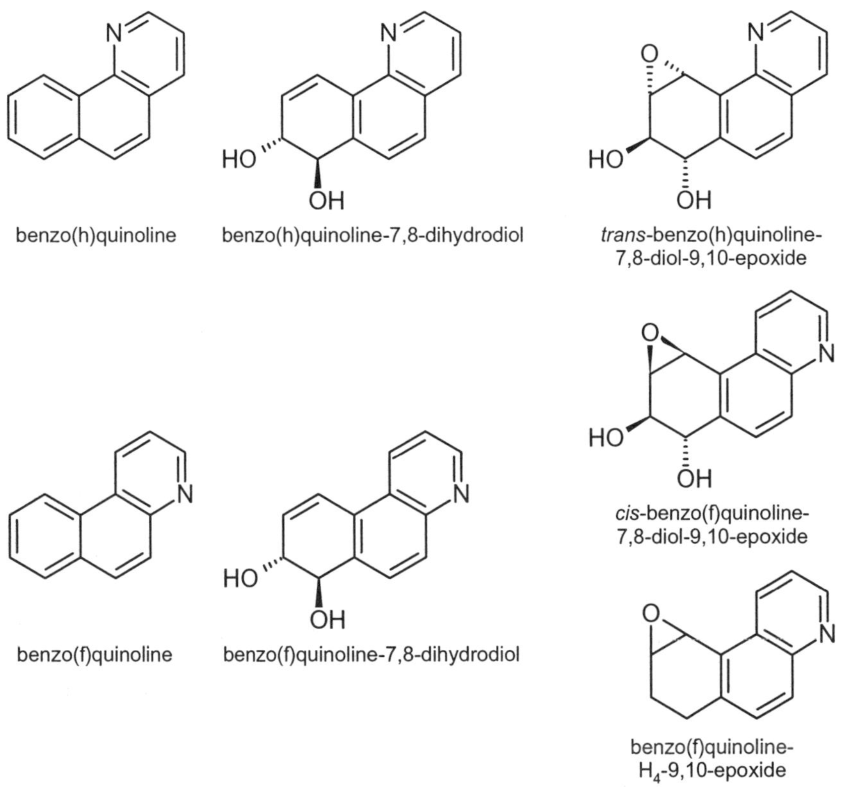

Fig. 13. Benzoquinoline metabolites that play a role in mutagenicity of these compounds (after Kumar et al. 1989).

D. Benzacridines

The general mechanism in mutagenicity of benzacridines appears to be similar to that of homocyclic PAHs, i.e., via diol epoxide intermediates, although benzacridines proved to be directly genotoxic in the Mutatox™ test (Bleeker et al. 1999b).

In tests with two different strains of *Salmonella* bacteria and Chinese hamster V79 cells, benz(*a*)acridine showed less mutagenic activity than benz(*c*)acridine (Wood et al. 1983). Similar to the mechanism in benzoquinolines, the bay region (diol) epoxides of these benzacridines were the most mutagenic derivatives. Furthermore, benz(*c*)acridine-H_4-1,2-epoxide and benz(*c*)acridine-3,4-diol-1,2-epoxide were more mutagenic than their respective benz(*a*)acridine isomers (Wood et al. 1983). Similar to the mutagenicity of benzoquinolines, this can be explained by the bay region nitrogen atom of benz(*c*)acridine that stabilizes the bay region epoxide more than would the K region nitrogen atom of benz(*a*)-acridine (Warshawsky 1992). In the Mutatox™ test a similar mechanism may induce the stronger genotoxic activity of benz(*a*)acridine compared to that of benz(*c*)acridine (Bleeker et al. 1999b), assuming that the nitrogen atom is the driving force behind the direct genotoxicity of these compounds.

Most benzacridines are regarded as noncarcinogenic in the mouse, but benz-(*c*)acridine has been shown to be a weak tumor initiator (Levin et al. 1983). Only 3 of the 12 derivatives tested showed higher tumor-initiating activity, *trans*-benz(*c*)acridine-3,4-diol-1,2-epoxide, benz(*c*)acridine-3,4-dihydrodiol, and 3,4-H_2-benz(*c*)acridine. In accordance with mutagenicity data this result suggests a bay region activation of benz(*c*)acridine toward an ultimate carcinogen. Consistent with these results are the results of a similar study of benz(*c*)acridine and these derivatives in newborn mice (Chang et al. 1984) and the observation that K region electron density is associated with higher carcinogenicities of benzacridines (Dipple et al. 1984).

E. Dibenzacridines

Dibenz(*a,h*)acridine, dibenz(*a,j*)acridine, and dibenz(*c,h*)acridine exhibit weak to moderate mutagenicity after metabolic activation (Bonin et al. 1989; Ho et al. 1981; Wood et al. 1986, 1989). Similar to the benzacridines and the benzoquinolines, dibenzacridines are initially metabolized to dihydrodiols, which are further metabolized to diol epoxides (Warshawsky 1992; Warshawsky et al. 1996; Wood et al. 1986, 1989). Other similarities with the smaller azaarenes are the stronger mutagenicity of bay region epoxide versus K region epoxides and of *trans*-epoxides versus *cis*-epoxides (Warshawsky et al. 1996; Wood et al. 1986, 1989).

Carcinogenicity of dibenzacridines proceeds through similar pathways as that of homocyclic analogues. The influence of the nitrogen atom on the electron density of the molecule is much smaller compared to the smaller azaarenes, resulting in biotransformation pathways similar to that of homocyclic analogues and thus similar active degradation products (Talaska et al. 1995; Warshawsky et al. 1994). In addition, a novel mechanism for DNA adduct formation by *N*-heterocycles was described, similar to that of dibenz(*a,h*)anthracene involving bis-dihydrodiol epoxides (Talaska et al. 1995). In another study (Warshawsky et al. 1994), dibenz(*a,j*)acridine induced skin tumors in mice, suggesting that DNA adducts may lead to carcinogenicity.

VI. Teratogenicity and Other Developmental Effects

Teratogenicity involves structural abnormalities in embryos. Effective exposure can occur either before conception (when one or both parents are exposed) or after conception, when either the mother or the developing embryo are exposed. Developing organisms can be particularly sensitive to toxicant exposure because they have not yet developed complete detoxification systems and because of ongoing organ differentiation and active growth. Teratogenicity is often referred to in terms of mammalian-type systems but can also be applied to other major taxonomic groups, such as insects or amphibians (*Xenopus laevis*).

Very few teratogenicity data are available for azaarenes, but in the early 1980s some studies on this subject were reported. Davis et al. (1981) and Dumont et al. (1979) demonstrated that both acridine and quinoline were teratogenic in the amphibian *Xenopus laevis*, and both compounds also showed teratogenicity for bass (*Micropterus salmoides*) and trout (*Oncorhynchus mykiss*) (Black et al. 1983). For the pouch snail (*Physa gyrina*) only quinoline was tested, but again it showed teratogenic effects (Millemann and Ehrenberg 1982). In humans, larger azaarenes, particularly, e.g., benzacridines, can cause functional birth defects such as the disease phenylketonuria by bringing a mutation into the phenylalanine hydroxylase gene (Ferguson and Denny 1991).

In contrast with the results mentioned here, in which azaarenes appeared to induce direct teratogenicity, Walton et al. (1983) held the ketone group responsible for teratogenic activity of the azaarene derivatives benz(*g*)isoquinoline-5,10-dione and benzo(*h*)quinoline-5,6-dione. These chemicals have been shown to induce a variety of physical abnormalities in developing cricket embryos at a range of doses (Walton et al. 1983). This study further showed the importance of the position of the nitrogen in the ring, as demonstrated by the absence of teratogenic effects of benzo(*g*)quinoline-5,10-dione in contrast to its teratogenic isomer benz(*g*)isoquinoline-5,10-dione.

One of the very few studies on other developmental effects of azaarenes, apart from teratogenicity, involved deformities of mouthparts in midge larvae (*Chironomus riparius*). Bleeker et al. (1999a) exposed the midge from the eggs onward to six different azaarenes. They studied fluctuating symmetry in one of the mouthparts of fourth-instar larvae, but in none of the exposures were significant differences from controls found, which may indicate that this hormone-regulated mechanism is not influenced by azaarenes, although concentrations tested were rather low.

VII. Comparing Azaarenes and Homocyclic PAHs
A. Comparing Toxicity

Azaarenes are suspected to be more toxic than homocyclic PAHs because the higher water solubility of the former (Pearlman et al. 1984) may increase their biological significance, resulting in a higher bioavailability and a higher toxicity. Comparing toxicity of azaarenes with that of homocyclic PAHs is difficult, because reported data on such comparisons are limited. The limited literature available, however, shows acridine toxicity being at least a factor of 5 lower than that of anthracene (Muñoz and Tarazona 1993; Wernersson and Dave 1997). Bioconcentration factor (BCF) values and LT_{50} values also indicate a higher toxicity of anthracene (Newsted and Giesy 1987). However, because both these compounds are phototoxic (see Section IV), the comparison between these compounds may not be representative for comparing toxicity of both groups of compounds. Furthermore, a comparison between homologue structures such as anthracene and acridine may be misleading because the nitrogen in acridine causes large differences in physicochemical properties between the compounds.

Like homocyclic PAHs, however, azaarenes exhibit baseline toxicity that can be predicted by log K_{ow}-based QSARs (Fig. 14). Any strong deviations from these relationships can usually be attributed to phototoxicity (acridine and benzacridines).

In biotransformation, the N-moiety of larger azaarenes, four rings and higher, becomes of less influence, resulting in transformation pathways similar to those of homocyclic PAHs. The smaller azaarenes in contrast can show strong differences in transformation pathways and thus in products formed (see Section II). Considering the toxic effects, however, differences are much smaller. Similar to homocyclic PAHs, biotransformation of azaarenes often leads to a reduction in narcotic activity of the compound, and also heterocyclic transformation products are usually more genotoxic (and carcinogenic) than their parent compounds.

Mutagenicity of azaarenes appears to proceed via similar pathways as those for homocyclic PAHs, resulting in similar active compounds. As for homocyclic

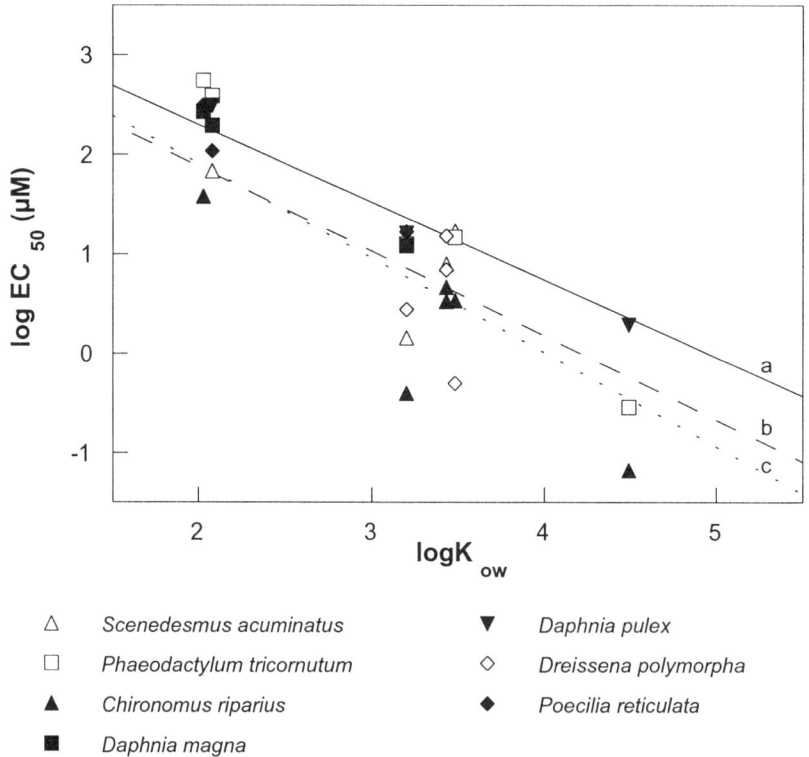

Fig. 14. Acute toxicity (data from Fig. 9) plotted versus log K_{ow} values (taken from Bleeker et al. 1998). a–c: relationships based on QSARs for narcotic chemicals (taken from Van Leeuwen et al. 1992); a: based on 96 hr LC_{50} for the copepod *Nitocra spinipes*; b: based on 96 hr LC_{50} for the fish *Pimephales promelas*; c: based on 48 hr LC_{50} for the daphnid *Daphnia magna*.

compounds, metabolic activation is generally needed for azaarenes to exhibit genotoxicity, although acridine has been proven to induce direct genotoxicity. Carcinogenicity of azaarenes is generally restricted to four-ringed and larger structures, and mechanisms leading to cancer are similar to those of homocyclic aromatics. Exception to this general pattern is quinoline, which has been shown to induce liver cancer.

In general it can be concluded that azaarene toxicity can be predicted quite well from models for homocyclic PAHs. The different metabolic pathways involved, however, may lead to additional, different types of toxicity, especially for the smaller (fewer than four rings) azaarenes.

B. Risk Assessment

In the Netherlands, risk assessment of PAHs is based on homocyclic compounds only, in particular a group of 10 representative PAHs (Slooff et al. 1989). This review, however, showed the ecotoxicological relevance of azaarenes, which is only one small group of heterocyclic compounds: at concentrations comparable to those in the field (0.1–1 µg/l; Van Genderen et al. 1994), toxicity of azaarenes has already been shown. This finding stresses the importance of incorporating heterocyclic compounds in risk assessment, especially when considering specific effects apart from baseline toxicity. The same holds for transformation products.

Another point of concern is the inappropriate inclusion of time dependency. Although this time dependency has been recognized, which has led to the calculation of acute-to-chronic ratios (Kenaga 1982; Länge et al. 1998; Mayer 1990; Roex et al. 2000), caution is necessary in using such ratios (see Section III).

Structure–activity relationships have been widely recognized as a useful tool in risk assessment. Such relationships, however, often focus on one particular biological endpoint while neglecting others. Yet, one compound can influence several endpoints, depending on exposure dose and duration, indicating that if risk assessment is solely based on one-endpoint quantitative structure–activity relationships (QSARs), the chance of missing other endpoints cannot be neglected. Therefore, it seems essential for a systematic analysis of effects of (N)PAHs to construct calculation routines for toxic action that provide effective explanations for overall effects. So far, this seems to be effective for mixed narcotic-phototoxic effects (Mekenyan et al. 1994a), but for other combinations of different types of toxicity no such SAR approaches have been used. To describe genotoxicity, the energy of the LUMO seems promising (Baeten et al. 1999; Tafazoli et al. 1998), but in chronic toxicity such parameters are still lacking, especially because acute-to-chronic ratios do not suffice for azaarenes (see Section III).

VIII. Conclusions

- Major metabolic products of the azaarenes appear to be ketones and mono- or dihydroxylated azaarenes.
- The metabolism of the larger azaarenes in vertebrates proceeds analogous to

homoaromatic PAH because in these larger systems the N-moiety is of less influence.
- In general, genotoxicity of azaarenes appears to follow similar pathways as homocyclic PAHs, often involving reactive epoxides.
- For both quinoline and acridine, differences in direct toxicity between species are small, suggesting essentially nonspecific activity.
- Large differences in toxicity between isomers demand the development of sophisticated, refined structure–activity relationships.
- The few data available on chronic toxicity indicate the limitations of acute-to-chronic ratios, suggesting that they should be used with great care.
- Photoenhanced toxicity has been shown predictable from electronic properties in both homocyclic PAHs as well as azaarenes. This knowledge is still neglected in risk assessment for both groups of compounds.
- Carcinogenicity of azaarenes has been observed in four-ringed and larger structures, and mechanisms leading to cancer in mammals are similar to those of homocyclic aromatics.
- The occasional record on teratogenic and other developmental effects makes it impossible to find patterns in these types of effects.
- Based on log K_{ow} models, azaarene toxicity does not differ from that of homocyclic PAHs. Yet, the higher water solubility of azaarenes enhances the chance of finding azaarene toxicity in the field, despite the lower concentrations compared to homocyclic PAHs.
- Data on chronic toxicity of (N)PAHs, including teratogenicity and other developmental effects, are a necessity in producing reliable acute-to-chronic ratios and therewith a better protection of the environment against long-term effects.
- Incorporating the role of heterocyclic compounds and metabolism appears to be a necessity for a reliable risk assessment for polycyclic aromatic compounds.

Summary

Heterocyclic compounds by far outnumber the homocyclic PAHs. In addition, they are often more soluble in water, which may imply a greater biological significance of these heterocycles. Yet, most research focuses on the homocyclics, based on the implicit assumption that the mostly higher concentration of the homocyclics rank these compounds as priority compounds. This review critically examines the available evidence and poses questions on the biological activity and environmental risk of one small group of heterocyclics, the azaarenes, which contain one nitrogen atom in one of the aromatic rings. In different sections, the biotransformation and different types of toxicity are discussed in comparison to those of homocyclic PAHs. The last section focuses on the implications for risk assessment of PAHs.

Two- and three-ringed azaarenes can be relatively easily transformed by bacteria, fungi, invertebrates, and vertebrates. The presence of the N-moiety in the smaller azaarenes leads to metabolic routes that partly differ from those of the homoaromatic analogues. Major metabolic products of the azaarenes appear to be ketones and mono- or dihydroxylated azaarenes. Microorganisms can further degrade these into multiple oxygen-containing compounds or they can open up the aza-containing aromatic ring and fully metabolize the products. Fungi and vertebrates were shown to produce the mutagenic dihydrodiol metabolites. The metabolism of the larger azaarenes in vertebrates proceeds analogous to homoaromatic PAH, because in these larger molecules the N-moiety has less influence. Transformation of the larger azaarenes by microorganisms proceeds much slower if occurring at all.

Direct toxicity data of azaarenes are mostly restricted to the effects of acridine and quinoline on a relatively small number of species. From this limited set it becomes clear that differences between species are relatively small. As with homocyclic PAHs, toxicity generally increases with increasing number of rings, and baseline toxicity models based on homocyclic PAHs do apply. Toxicity differences between isomers indicate that azaarene toxicity cannot be explained by molecular size-related parameters alone, indicating that electronic forces may be important as well. Considering chronic toxicity it becomes clear that the often-used acute-to-chronic-ratios often underestimate specific chronic toxicity, even within the very limited set of chronic data available.

In contrast with homocyclic PAHs, photodegradation of azaarenes shows the same degradation products as biological transformation involving monooxygenases. In general, as for homocyclic PAHs, the degree of phototoxicity is related to the UV absorption characteristics of the azaarenes, which makes it possible to apply the QSAR models developed for homocyclic PAHs to azaarenes as well. Recent research on algae showed that UV-A is the main cause of photoenhanced toxicity. Together with the fact that in the water column UV-B is almost absent, this clearly demonstrates the relevance of phototoxicity in the field.

Mutagenicity of azaarenes generally proceeds through similar pathways as in homocyclic PAHs, with bay region diol epoxides as major genotoxic metabolites. The N-moiety can, however, result in differences in genotoxic activities between isomers. Carcinogenicity of azaarenes in mammals is generally restricted to four-ringed and larger structures, and mechanisms leading to cancer are similar to those of homocyclic aromatics. An exception to this general pattern is quinoline, which has been shown to induce liver cancer.

The present risk assessment for PAHs is solely based on homocyclic PAHs. Yet, from the present review it becomes clear that this approach fails to protect against a vast number of heterocyclic compounds and biotransformation products that may exhibit stronger or other toxic effects than their homocyclic analogues. Therefore, incorporating the role of heterocyclic compounds and their metabolism appears to be a necessity for a reliable risk assessment for polycyclic aromatic compounds. In addition, reliable long-term protection against PAHs demands data on chronic toxicity, including teratogenicity, both for homocyclic as for heterocyclic compounds.

References

Adams EA, LaVoie EJ, Hoffmann D (1983) Mutagenicity and metabolism of azaphenanthrenenes. In: Cooke M (ed) Polynuclear Aromatic Hydrocarbons. Battelle Press, Columbus, OH, pp 73–84.

Adams J, Giam C-S (1984) Polynuclear azaarenes in wood preservative wastewater. Environ Sci Technol 18:391–394.

Adema DMM, Canton JH, Slooff W, Hanstveit AO (1981) Onderzoek naar een geschikte combinatie toetsmethoden ter bepaling van de aquatische toxiciteit van milieugevaarlijke stoffen. RIV/RID/TNO report CL 81/100; RIV 627905 001; RID CBH 81/11. Ministerie van Volksgezondheid en Milieuhygiëne, The Hague, The Netherlands.

Aislabie J, Rothenburger S, Atlas RM (1989) Isolation of microorganisms capable of degrading isoquinoline under aerobic conditions. Appl Environ Microbiol 55:3247–3249.

Aislabie J, Bej AK, Hurst H, Rothenburger S, Atlas RM (1990) Microbial-degradation of quinoline and methylquinolines. Appl Environ Microbiol 56:345–351.

Ames BN, Durston WE, Yamasaki E, Lee FD (1973a) Carcinogens are mutagens: a simple test system combining liver homogenates for activation and bacteria for detection. Proc Natl Acad Sci USA 70:2281–2285.

Ames BN, Lee FD, Durston WE (1973b) An improved bacterial test system for the detection and classification of mutagens and carcinogens. Proc Natl Acad Sci USA 70:782–786.

Ankley GT, Collyard SA, Monson PD, Kosian PA (1994) Influence of ultraviolet light on the toxicity of sediments contaminated with polycyclic aromatic hydrocarbons. Environ Toxicol Chem 13:1791–1796.

Ankley GT, Erickson RJ, Phipps GL, Mattson VR, Kosian PA, Sheedy BR, Cox JS (1995) Effects of light intensity on the phototoxicity of fluoranthene to a benthic macroinvertebrate. Environ Sci Technol 29:2828–2833.

Ankley GT, Erickson RJ, Sheedy BR, Kosian PA, Mattson VR, Cox JS (1997) Evaluation of models for predicting the phototoxic potency of polycyclic aromatic hydrocarbons. Aquat Toxicol 37:37–50.

Appel KE, Fürstenberger G, Hapke HJ, Hecker E, Hildebrandt AG, Koransky W, Marks F, Neumann HG, Ohnesorge FK, Schulte-Hermann R (1990) Chemical Cancerogenesis: definitions of frequently used terms. J Cancer Res Clin Oncol 116:232–236.

Arfsten DP, Schaeffer DJ, Mulveny DC (1996) The effects of near ultraviolet radiation on the toxic effects of polycyclic aromatic hydrocarbons in animals and plants: a review. Ecotoxicol Environ Saf 33:1–24.

Ashok BT, Saxena S (1995) Biodegradation of polycyclic aromatic-hydrocarbons—a review. J Sci Ind Res 54:443–451.

Baek SO, Field RA, Goldstone ME, Kirk PW, Lester JN, Perry R (1991) A review of atmospheric polycyclic aromatic-hydrocarbons—sources, fate and behavior. Water Air Soil Pollut 60:279–300.

Baeten A, Tafazoli M, Kirsch-Volders M, Geerlings P (1999) Use of the HSAB principle in quantitative structure-activity relationships in toxicological research: Application to the genotoxicity of chlorinated hydrocarbons. Int J Quantum Chem 74:351–355.

Baumard P, Budzinski H, Garrigues P (1998) Polycyclic aromatic hydrocarbons in sediments and mussels of the Western Mediterranean Sea. Environ Toxicol Chem 17:765–776.

Bean RM, Dauble DD, Thomas BL, Hanf RW Jr, Chess EK (1985) Uptake and biotransformation of quinoline by rainbow trout. Aquat Toxicol 7:221–239.

Becker WM, Reece JB, Poenie MF (1996) The World of the Cell, 3rd Ed. Benjamin/Cummings, Menlo Park, CA.

Benson JM, Royer RE, Hill JO (1982) Metabolism of phenanthridine, an aza-arene present in low Btu gasifier effluents. In: Cooke M, Dennis AJ (eds) Polynuclear Aromatic Hydrocarbons: Physical and Biological Chemistry. Battelle Press, Columbus, OH, pp 103–108.

Benson JM, Royer RE, Galvin JB, Shimizu RW (1983) Metabolism of phenanthridine to phenanthridone by rat lung and liver microsomes after induction with benzo[a]pyrene and aroclor. Toxicol Appl Pharmacol 68:36–42.

Bernges F, Zeller WJ (1996) Combination effects of poly(ADP-ribose) polymerase inhibitors and DNA-damaging agents in ovarian tumor cell lines: with special reference to cisplatin. J Cancer Res Clin Oncol 122:665–670.

Black JA, Birge WJ, Westerman AG, Francis PC (1983) Comparative aquatic toxicology of aromatic hydrocarbons. Fundam Appl Toxicol 3:353–358.

Blaylock BC, Frank ML, McCarthy JF (1985) Comparative toxicity of copper and acridine to fish, *Daphnia* and algae. Environ Toxicol Chem 4:63–71.

Bleeker EAJ, Admiraal W, Buckert-de Jong MC, Van der Geest HG, Kraak MHS, Van Vlaardingen P, De Voogt P, Wind E (1996) NPAHs in Dutch marine and freshwater sediments and toxic, mutagenic and teratogenic effects on aquatic invertebrates. ARISE report. University of Amsterdam, Amsterdam, The Netherlands.

Bleeker EAJ, Van der Geest HG, Kraak MHS, De Voogt P, Admiraal W (1998) Comparative ecotoxicity of NPAHs to larvae of the midge *Chironomus riparius*. Aquat Toxicol 41:51–62.

Bleeker EAJ, Leslie HA, Groenendijk D, Plans M, Admiraal W (1999a) Effects of exposure to azaarenes on emergence and mouthpart development in the midge *Chironomus riparius* (Diptera: Chironomidae). Environ Toxicol Chem 18:1829–1834.

Bleeker EAJ, Van der Geest HG, Klamer HJC, De Voogt P, Wind E, Kraak MHS (1999b) Toxic and genotoxic effects of azaarenes: isomers and metabolites. Polycyclic Arom Comp 13:191–203.

Bleeker EAJ, Noor L, Kraak MHS, De Voogt P, Admiraal W (2001) Comparative metabolism of phenanthridine by carp (*Cyprinus carpio*) and midge larvae (*Chironomus riparius*). Environ Pollut 112:11–17.

Blumer M, Dorsey T, Sass J (1977) Azaarenes in recent marine sediments. Science 195:283–285.

Boese BL, Lamberson JO, Swartz RC, Ozretich R, Cole F (1998) Photoinduced toxicity of PAHs and alkylated PAHs to a marine infaunal amphipod (*Rhepoxynius abronius*). Arch Environ Contam Toxicol 34:235–240.

Bollag J-M, Kaiser J-P (1991) The transformation of heterocyclic aromatic compounds and their derivates under anaerobic conditions. Crit Rev Environ Control 21:297–329.

Bonin AM, Rosario CA, Duke CC, Baker RSU, Ryan AJ, Holder GM (1989) The mutagenicity of dibenz[a,j]acridine, some metabolites and other derivatives in bacteria and mammalian cells. Carcinogenesis 10:1079–1084.

Boyd DR, Sharma ND, Dorrity MRJ, Hand MV, McMordie RAS, Malone JF, Porter HP, Dalton H, Chima J, Sheldrake GN (1993) Structure and stereochemistry of *cis*-dihydro diol and phenol metabolites of bicyclic azaarenes from *Pseudomonas putida* Uv4. J Chem Soc Perkin Trans 1:1065–1071.

Cassidy RA, Birge WJ, Black JA (1988) Biodegradation of three azaarene congeners in river water. Environ Toxicol Chem 7:99–105.

Chang RL, Levin W, Wood AW, Kumar S, Yagi H, Jerina DM, Lehr RE, Conney AH (1984) Tumorigenicity of dihydrodiols and diol-epoxides of benz[c]acridine in newborn mice. Cancer Res 44:5161–5164.

Chen J, Wang L, Lu G, Zhao T (1997) Quantitative structure-activity relationship studies of selected heterocyclic nitrogen compounds. Bull Environ Contam Toxicol 58:372–379.

Cooney JD, Gehrs CW (1984) Effects of temperature, feeding and acridine on development and mortality of eggs and nauplii of *Diaptomus clavipes* Schacht. Aquat Toxicol 5:197–209.

Cowan DA, Damani LA, Gorrod JW (1978) Metabolic N-oxidation of 3-substituted pyridines: identifiation of products by mass spectrometry. Biomed Mass Spectrom 5: 551–556.

Cushman RM, McKamey MI (1981) A *Chironomus tentans* bioassay for testing synthetic fuel products and effluents, with data on acridine and quinoline. Bull Environ Contam Toxicol 26:601–605.

David B, Boule P (1993) Phototransformation of hydrophobic pollutants in aqueous medium. 1. PAHs adsorbed on silica. Chemosphere 26:1617–1630.

Davis KR, Schultz TW, Dumont JN (1981) Toxic and teratogenic effects of selected aromatic amines on embryos of the amphibian *Xenopus laevis*. Arch Environ Contam Toxicol 10:371–391.

De Maagd PG-J, Tonkes M (2000) Selection of genotoxicity tests for risk assessment of effluents. Environ Toxicol 15:81–90.

De Voogt P, Wegener JWM, Klamer JC, Van Zijl GA, Govers H (1988) Prediction of environmental fate and effects of heteroatomic polycyclic aromatics by QSARs: the position of n-octanol/water partition coefficients. Biomed Environ Sci 1:194–209.

De Voogt P, Bleeker EAJ, Van Vlaardingen PLA, Fernández A, Slobodník J, Wever H, Kraak MHS (1999) Formation and identification of azaarene transformation products from aquatic invertebrate and algal metabolism. J Chromatgr B 724:265–274.

Dijkman NA, Van Vlaardingen PLA, Admiraal WA (1997) Biological variation in sensitivity to N-heterocyclic PAHs; effects of acridine on seven species of microalgae. Environ Pollut 95:121–126.

Dipple A, Moschel RC, Bigger CAH (1984) Polynuclear aromatic carcinogens. In: Searle CE (ed) Chemical Carcinogens, Vol. 1, 2nd Ed. (ACS Monographs, Vol. 182.) American Chemical Society, Washington, DC, pp 41–163.

Dumont JN, Schultz TW, Jones RD (1979) Toxicity and teratogenicity of aromatic amines to *Xenopus laevis*. Bull Environ Contam Toxicol 22:159–166.

Edler B, Zwiener C, Frimmel FH (1997) Particle beam LC/MS screening of polar compounds of contaminated ground water samples from a former gas plant. Fresenius J Anal Chem 359:288–292.

Ferguson LR, Denny WA (1991) The genetic toxicology of acridines. Mutat Res 258: 123–160.

Foote CS (1991) Definition of type I and type Ii photosensitized oxidation. Photochem Photobiol 54:659.

Furlong ET, Carpenter R (1982) Azaarenes in Puget Sound sediments. Geochim Cosmochim Acta 46:1385–1396.

Gala WR, Giesy JP (1993) Using the carotenoid biosynthesis inhibiting herbicide, fluridone, to investigate the ability of carotenoid-pigments to protect algae from the photoinduced toxicity of anthracene. Aquat Toxicol 27:61–70.

Gala WR, Giesy JP (1994) Flow cytometric determination of the photoinduced toxicity of anthracene to the green alga *Selenastrum capricornutum*. Environ Toxicol Chem 13:831–840.

Giddings JM (1979) Acute toxicity to *Selenastrum capricornutum* of aromatic compounds from coal conversion. Bull Environ Contam Toxicol 23:360–364.

Gieg LM, Otter A, Fedorak PM (1996) Carbazole degradation by *Pseudomonas* sp. LD2: metabolic characteristics and the identification of some metabolites. Environ Sci Technol 30:575–585.

Gill JH, Duke CC, Roasario CA, Ryan AJ, Holder GM (1986) Dibenz[a,j]acridine metabolism: identification of in virtro products formed by liver microsomes from 3-methylcholanthrene-pretreated rats. Carcinogenesis 7:1371–1378.

Gissel-Nielsen G, Nielsen T (1996) Phytotoxicity of acridine, an important representative of a group of tar and creosote contaminants, N-PAC compounds. Polycyclic Arom Comp 8:243–249.

Grosser RJ, Warshawsky D, Vestal JR (1995) Mineralization of polycyclic and N-heterocyclic aromatic compounds in hydrocarbon-contaminated soils. Environ Toxicol Chem 14:375–382.

Hansch C, Leo A, Hoekman D (1995) Exploring QSAR: Hydrophobic, Electronic, and Steric constants. (*ACS* Professional Reference Book) American Chemical Society. Washington, DC, USA.

Helweg C, Nielsen T, Hansen PE (1997) Determination of octanol-water partition coefficients of polar polycyclic aromatic compounds (N-PAC) by high performance liquid chromatography. Chemosphere 34:1673–1684.

Hermens J, Canton H, Steyger N, Wegman R (1984) Joint effects of a mixture of 14 chemicals on mortality and inhibition of reproduction of *Daphnia magna*. Aquat Toxicol 5:315–322.

Hirao K, Shinohara Y, Tsuda H, Fukushima S, Takahashi M, Ito N (1976) Carcinogenic activity of quinoline on rat liver. Cancer Res 36:329–335.

Ho C-H, Clark BR, Guerin MR, Barkenbus BD, Rao TK, Epler JL (1981) Analytical and biological analyses of test materials from the synthetic fuel technologies: IV. Studies of chemical structure-mutagenic activity relationships of aromatic nitrogen compounds relevant to synfuels. Mutat Res 85:335–345.

Holst LL, Giesy JP (1989) Chronic effects of the photoenhanced toxicity of anthracene on *Daphnia magna* reproduction. Environ Toxicol Chem 8:933–942.

Howard PC, Hecht SS, Beland FA (1991) Nitroarenes: Occurence, Metabolism, and Biological impact, Vol. 40. Plenum, New York.

Huang X-D, Dixon DG, Greenberg BM (1993) Impacts of UV radiation and photomodification on the toxicity of PAHs to the higher plant *Lemna gibba* (duckweed). Environ Toxicol Chem 12:1067–1077.

Huang X-D, Dixon DG, Greenberg BM (1995) Increased polycyclic aromatic hydrocarbon toxicity following their photomodification in natural sunlight: impacts on the duckweed *Lemna gibba* L.G-3. Ecotoxicol Environ Saf 32:194–200.

Huang X-D, Krylov SN, Ren L, McConkey BJ, Dixon DG, Greenberg BM (1997a) Mechanistic quantitative structure-activity relationship model for the photoinduced toxicity of polycyclic aromatic hydrocarbons: II. An empirical model for the toxicity of 16 polycyclic aromatic hydrocarbons to the duckweed *Lemna gibba* L. G-3. Environ Toxicol Chem 16:2296–2303.

Huang X-D, McConkey BJ, Babu TS, Greenberg BM (1997b) Mechanisms of photoinduced toxicity of photomodified anthracene to plants: inhibition of photosynthesis in the aquatic higher plant *Lemna gibba* (duckweed). Environ Toxicol Chem 16:1707–1715.

Jacob J, Schmoldt A, Raab G, Kohbrok W, Grimmer G (1983) Rat liver and lung microsomal metabolism of benz[a]- and benz[c]acridine. In: Cooke M (ed) Polynuclear Aromatic Hydrocarbons. Battelle Press, Columbus, OH, pp 637–647.

Jerina DM, Lehr RE (1977) The bay region theory: a quantum mechanical approach to aromatic hydrocarbon-induced carcinogenicity. In: Ullrich V, Roots I, Hildebrant AG, Estabrook RW, Conney AH (eds) Microsomes and Drug Oxidations. Pergamon Press, Oxford, pp 709–720.

Johansen SS, Arvin E, Mosbaek H, Hansen AB (1997a) Degradation pathway of quinolines in a biofilm system under denitrifying conditions. Environ Toxicol Chem 16: 1821–1828.

Johansen SS, Licht D, Arvin E, Mosbaek H, Hansen AB (1997b) Metabolic pathways of quinoline, indole and their methylated analogs by *Desulfobacterium indolicum* (DSM 3383). Appl Microbiol Biotechnol 47:292–300.

Johnson BT (1992) An evaluation of a genotoxicity assay with liver S9 for activation and luminescent bacteria for detection. Environ Toxicol Chem 11:473–480.

Jones KC, Stratford JA, Tidridge P, Waterhouse KS, Johnston AE (1989) Polynuclear aromatic hydrocarbons in an agricultural soil: long-term changes in profile distribution. Environ Pollut 56:337–351.

Kagan J, Kagan ED, Kagan IA, Kagan PA, Quigley S (1985) The phototoxicity of noncarcinogenic polycyclic aromatic hydrocarbons in aquatic organisms. Chemosphere 14:1829–1834.

Kagan J, Stokes A, Gong H-H, Tuveson RW (1987) Light-dependent cytotoxicity of fluoranthene: oxygen-dependent membrane damage. Chemosphere 16:2417–2422.

Kaiser J-P, Feng Y, Bollag J-M (1996) Microbial metabolism of pyridine, quinoline, acridine, and their derivatives under aerobic and anaerobic conditions. Microbiol Rev 60:483–498.

Kalf DF, Crommentuijn GH, Posthumus R, Van de Plassche EJ (1995) Integrated environmental quality objectives for polycyclic aromatic hydrocarbons (PAHs). RIVM report 679101 018. National Institute of Public Health and Environmental Protection (RIVM), Bilthoven, The Netherlands.

Kamath AV, Vaidyanathan CS (1990) New pathway for the biodegradation of indole in *Aspergillus niger*. Appl Environ Microbiol 56:275–280.

Kandaswami C, Kumar S, Dubey SK, Sikka HC (1987) Metabolism of benzo[f]quinoline by rat liver microsomes. Carcinogenesis 8:1861–1866.

Kawamura K, Suzuki I, Fujii Y, Watanabe O (1994) Ice core record of polycyclic aromatic hydrocarbons over the past 400 years. Naturwissenschaften 81:502–505.

Kenaga EE (1982) Predictability of chronic toxicity from acute toxicity of chemicals in fish and aquatic invertebrates. Environ Toxicol Chem 1:347–358.

Khalili NR, Scheff PA, Holsen TM (1995) PAH source fingerprints for coke ovens, diesel and gasoline engines, highway tunnels, and wood combustion emissions. Atmos Environ 29:533–542.

Kilbane JJ, Ranganathan R, Cleveland L, Kayser KJ, Ribiero C, Linhares MM (2000) Selective removal of nitrogen from quinoline and petroleum by *Pseudomonas ayucida* IGTN9m. Appl Environ Microbiol 66:688–693.

Klaassen CD, Amdur MO, Doull J (1996) Casarett and Doull's Toxicology: The Basic Science of Poisons, 5th Ed. Pergamon, Oxford.

Knezovich JP, Bishop DJ, Kulp TJ, Grbic-Galic D, Dewitt J (1990) Anaerobic microbial degradation of acridine and the application of remote fiber spectroscopy to monitor the transformation process. Environ Toxicol Chem 9:1235–1243.

Kochany J, Maguire RJ (1994) Abiotic transformations of polynuclear aromatic hydrocarbons and polynuclear aromatic nitrogen heterocycles in aquatic environments. Sci Total Environ 144:17–31.

Könemann H (1981) Quantitative structure-activity relationships in fish toxicity studies. Part 1: Relationship for 50 industrial pollutants. Toxicology 19:209–221.

Kozin IS, Larsen OFA, De Voogt P, Gooijer C, Velthorst NH (1997) Isomer-specific detection of azaarenes in environmental samples by Shpol'skii luminescence spectroscopy. Anal Chim Acta 354:181–187.

Kraak MHS, Ainscough C, Fernández A, Van Vlaardingen PLA, De Voogt P, Admiraal W (1997a) Short-term and chronic exposure of the zebra mussel (*Dreissena polymorpha*) to acridine: effects and metabolism. Aquat Toxicol 37:9–20.

Kraak MHS, Wijnands P, Govers HAJ, Admiraal W, De Voogt P (1997b) Structural-based differences in ecotoxicity of benzoquinoline isomers to the zebra mussel (*Dreissena polymorpha*). Environ Toxicol Chem 16:2158–2163.

Kropp KG, Fedorak PM (1998) A review of the occurrence, toxicity, and biodegradation of condensed thiophenes found in petroleum. Can J Microbiol 44:605–622.

Krylov SN, Huang X-D, Zeiler LF, Dixon DG, Greenberg BM (1997) Mechanistic quantitative structure-activity relationship model for the photoinduced toxicity of polycyclic aromatic hydrocarbons: I. Physical model based on chemical kinetics in a two-compartment system. Environ Toxicol Chem 16:2283–2295.

Kuhn EP, Suflita JM (1989) Microbial degradation of nitrogen, oxygen and sulfur heterocyclic compounds under anaerobic conditions. Environ Toxicol Chem 8:1149–1158.

Kühn R, Pattard M, Pernak KD, Winter A (1989) Results of the harmful effects of water pollutants to *Daphnia magna* in the 21 day reproduction test. Water Res 23:501–510.

Kumar S, Sikka HC, Dubey SK, Czech A, Geddie N, Wange C-X, LaVoie EJ (1989) Mutagenicity and tumorigenicity of dihydrodiols, diol epoxides, and other derivatives of benzo[*f*]quinoline and benzo[*h*]quinoline. Cancer Res 49:20–24.

Länge R, Hutchinson TH, Scholz N, Solbé J (1998) Analysis of the ECETOC aquatic toxicity (EAT) database II: comparison of acute to chronic ratios for various aquatic organisms and chemicals substances. Chemosphere 36:115–127.

Larson RA, Berenbaum MR (1988) Environmental phototoxicity. Environ Sci Technol 22:354–360.

LaVoie EJ, Adams EA, Shigematsu A, Hoffmann D (1983) On the metabolism of quinoline and isoquinoline: possible molecular basis for differences in biological activities. Carcinogenesis 4:1169–1173.

LaVoie EJ, Adams EA, Shigematsu A, Hoffmann D (1985) Metabolites of phenanthridine formed by rat liver homogenate. Drug Metab Dispos 13:71–75.

LaVoie EJ, Shigematsu A, Adams EA, Geddie NG, Rice JE (1988) Quinolines and benzoquinolines: studies related to their metabolism, mutagenicity, tumor-initiating activity, and carcinogenicity. In: Cooke M, Dennis AJ (eds) Polynuclear Aromatic Hydrocarbons: A Decade of Progress. Battelle Press, Columbus, OH, pp 503–518.

Lee RF (1988) Possible linkages between mixed function oxygenase systems, steroid metabolism, reproduction, molting, and pollution in aquatic animals. In: Evans MS (ed) Advances in Environmental Science and Technology. (Toxic Contaminants and Ecosystem Health: A Great Lakes Focus, Vol. 21.) Wiley, New York, pp 201–213.

Leifer A (1988) The Kinetics of Environmental Aquatic Photochemistry. American Chemical Society, Washington, DC.

Levin W, Wood AW, Chang RL, Kumar S, Yagi H, Jerina DM, Lehr RE, Conney AH (1983) Tumor-initiating activity of benz[c]acridine and twelve of its derivatives on mouse skin. Cancer Res 43:4625–4628.

Liu S-M, Jones WJ, Rogers JE (1994a) Influence of redox potential on the anaerobic biotransformation of nitrogen-heterocyclic compounds in anoxic freshwater sediments. Appl Microbiol Biotechnol 41:717–724.

Liu S-M, Jones WJ, Rogers JE (1994b) Biotransformation of quinoline and methylquinolines in anoxic freshwater sediment. Biodegradation 5:113–116.

Livingstone DR (1998) The fate of organic xenobiotics in aquatic ecosystems: quantitative and qualitative differences in biotransformation by invertebrates and fish. Comp Biochem Physiol A 120:43–49.

Livingstone DR, Farrar SV (1984) Tissue and subcellular distribution of enzyme activities of mixed-function oxygenase and benzo[*a*]pyrene metabolism in the common mussel *Mytilus edulis* L. Sci Total Environ 39:209–235.

Mallakin A, McConkey BJ, Miao G, McKibben B, Snieckus V, Dixon DG, Greenberg BM (1999) Impacts of structural photomodification on the toxicity of environmental contaminants: anthracene photooxidation products. Ecotoxicol Environ Saf B 43:204–212.

Marwood CA, Smith REH, Solomon KR, Charlton MN, Greenberg BM (1999) Intact and photomodified polycyclic aromatic hydrocarbons inhibit photosynthesis in natural assemblages of Lake Erie phytoplankton exposed to solar radiation. Ecotoxicol Environ Saf 44:322–327.

Mayer FL (1990) Predicting chronic lethality of chemicals to fishes from acute toxicity test data. EPA/600/X-90/147. U.S. Environmental Protection Agency, Environmental Research Laboratory, Gulf Breeze, FL.

McConkey BJ, Duxbury CL, Dixon DG, Greenberg BM (1997) Toxicity of a PAH photooxidation product to the bacteria *Photobacterium phosphoreum* and the duckweed *Lemna gibba*: effects of phenanthrene and its primary photoproduct, phenanthrenequinone. Environ Toxicol Chem 16:892–899.

McMurtrey KD, Knight TJ (1984) Metabolism of acridine by rat-liver enzymes. Mutat Res 140:7–11.

Mekenyan OG, Ankley GT, Veith GD, Call DJ (1994a) QSARs for photoinduced toxicity: I. Acute lethality of polycyclic aromatic hydrocarbons to *Daphnia magna*. Chemosphere 28:567–582.

Mekenyan OG, Ankley GT, Veith GD, Call DJ (1994b) QSAR estimates of excited states and photoinduced acute toxicity of polycyclic aromatic hydrocarbons. SAR QSAR Environ Res 2:237–247.

Michael JP (2000) Quinoline, quinazoline and acridone alkaloids. Nat Prod Rep 17:603–620.

Mill T, Mabey WR, Lan BY, Baraze A (1981) Photolysis of polycyclic aromatic hydrocarbons in water. Chemosphere 10:1281–1290.

Millemann RE, Ehrenberg DS (1982) Chronic toxicity of the azaarene quinoline, a synthetic fuel component, to the pond snail *Physa gyrina*. Environ Technol Lett 3:193–198.

Millemann RE, Birge WJ, Black JA, Cushman RM, Daniels KL, Franco PJ, Giddings

JM, McCarthy JF, Stewart AJ (1984) Comparative acute toxicity to aquatic organisms of components of coal-derived synthetic fuels. Trans Am Fish Soc 113:74–85.

Moir D, Poon R, Yagminas A, Park G, Viau A, Valli VE, Chu I (1997) The subchronic toxicity of acridine in the rat. J Environ Sci Health B 32:545–564.

Monson PD, Ankley GT, Kosian PA (1995) Phototoxic response of *Lumbriculus variegatus* to sediments contaminated by polycyclic aromatic hydrocarbons. Environ Toxicol Chem 14:891–894.

Müller MB, Zwiener C, Frimmel FH (1999) Sample cleanup and reversed-phase high-performance liquid chromatographic analysis of polar aromatic compounds in groundwater samples from a former gas plant. J Chromatogr A 862:137–145.

Muncnerova D, Augustin J (1994) Fungal metabolism and detoxification of polycyclic aromatic hydrocarbons: a review. Bioresour Technol 48:97–106.

Muñoz MJ, Tarazona JV (1993) Synergistic effect of two- and four-component combinations of the polycyclic aromatic hydrocarbons: phenanthrene, anthracen, naphtalene and acenaphthene on *Daphnia magna*. Bull Environ Contam Toxicol 50:363–368.

Neff JM (1985) Polycyclic aromatic hydrocarbons. In: Rand GM, Petrocelli SR (eds) Fundamentals of Aquatic Toxicology: Methods and Applications. Hemisphere, Washington, DC.

Newsted JL, Giesy JP (1987) Predictive models for photoinduced acute toxicity of polycyclic aromatic hydrocarbons to *Daphnia magna*, Strauss (Cladocera, Crustacea). Environ Toxicol Chem 6:445–461.

Nielsen T, Siigur K, Helweg C, Jørgensen O, Hansen P, Kirso U (1997) Sorption of polycyclic aromatic compounds to humic acid as studied by high-performance liquid chromatography. Environ Sci Technol 31:1102–1108.

Onodera K, Furusawa G, Kojima M, Tsuchiya M, Aihara S, Akaba R, Sakuragi H, Tokumaru K (1985) Mechanistic considerations on photoreaction of organic compounds via excitation of contact charge transfer complexes with oxygen. Tetrahedron 41:2215–2220.

Oris JT, Giesy JP (1985) The photoenhanced toxicity of anthracene to juvenile sunfish (*Lepomis* spp.). Aquat Toxicol 6:133–146.

Oris JT, Giesy JP (1986) Photoinduced toxicity of anthracene to juvenile bluegill sunfish (*Lepomis macrochirus* Rafinesque): photoperiod effects and predictive hazard evaluation. Environ Toxicol Chem 5:761–768.

Oris JT, Giesy JP Jr (1987) The photo-induced toxicity of polycyclic aromatic hydrocarbons to larvae of the fathead minnow (*Pimephales promelas*). Chemosphere 16:1395–1404.

Osborne PJ, Preston MR, Chen HY (1997) Azaarenes in sediments, suspended particles and aerosol associated with the River Mersey estuary. Mar Chem 58:73–83.

Oshiro Y, Sato S, Kurahashi N, Tanaka T, Kikuchi T, Tottori K, Uwahodo Y, Nishi T (1998) Novel antipsychotic agents with dopamine autoreceptor agonist properties: synthesis and pharmacology of 7-[4-(4-phenyl-1-piperazinyl)butoxy]-3,4-dihydro-2(1*H*)-quinoline derivatives. J Med Chem 41:658–667.

Palmer CM, Maloney TE (1955) Preliminary screening for potential algicides. Ohio J Sci 55:1–8.

Parkhurst BR, Bradshaw AS, Forte JL, Wright GP (1981) The chronic toxicity of *Daphnia magna* of acridine, a representative azaarene present in synthetic fossil fuel products and wastewaters. Environ Pollut Ser A 24:21–30.

Pearlman RS, Yalkowsky SH, Banerjee S (1984) Water solubility of polynuclear aromatic and heteroaromatic compounds. J Phys Chem Ref Data 13:555–562.

Pelletier MC, Burgess RM, Ho KT, Kuhn A, McKinney RA, Ryba SA (1997) Phototoxicity of individual polycyclic aromatic hydrocarbons and petroleum to marine invertebrate larvae and juveniles. Environ Toxicol Chem 16:2190–2199.

Pereira WE, Rostad CE, Garbarino JR, Hult MF (1983) Groundwater contamination by organic bases derived from coal-tar wastes. Environ Toxicol Chem 2:283–294.

Pereira WE, Rostad CE, Updegraff DM, Bennett JL (1987) Fate and movement of azaarenes and their anaerobic biotransformation products in an aquifer contaminated by wood-treatment chemicals. Environ Toxicol Chem 6:163–176.

Picel KC, Simmons MS, Stamoudis VC (1987) Sunlight photolysis of selected indoles and carbazole in aqueous coal-oil systems. In: Zika RG, Cooper WJ (eds) Photochemistry of Environmental Aquatic Systems. (ACS Symposium Series, Vol. 327.) American Chemical Society, Washington, DC, pp 44–60.

Rappeneau S, Baeza-Squiban A, Jeulin C, Marano F (2000) Protection from cytotoxic effects induced by the nitrogen mustard mechlorethamine on human bronchial epithelial cells in vitro. Toxicol Sci 54:212–221.

Ren L, Huang X-D, McConkey BJ, Dixon DG, Greenberg BM (1994) Photoinduced toxicity of three polycyclic aromatic hydrocarbons (fluoranthene, pyrene, and naphthalene) to the duckweed *Lemna gibba* L. G-3. Ecotoxicol Environ Saf 28:160–171.

Richardson DS, Allen PD, Kelsey SM, Newland AC (1999) Effects of PARP inhibition on drug and Fas-induced apoptosis in leukaemic cells. In: Drug Resistance in Leukemia and Lymphoma III. (Advances in Experimental Medicine and Biology, Vol. 457.) Kluwer Academic/Plenum, New York, pp 267–279.

Roex EWM, Van Gestel CAM, Van Wezel AP, Van Straalen NM (2000) Ratios between acute and chronic toxicity and effects on population growth rates in relation to toxiant mode of action. Environ Toxicol Chem 19:685–693.

Sanders G, Jones KC, Hamilton-Taylor J, Dörr H (1993) Concentrations and deposition fluxes of polynuclear aromatic hydrocarbons and heavy metals in the dated sediments of a rural English lake. Environ Toxicol Chem 12:1567–1581.

Santodonato J (1997) Review of the estrogenic and antiestrogenic activity of polycyclic aromatic hydrocarbons: relationship to carcinogenicity. Chemosphere 34:835–848.

Santodonato J, Howard PH (1981) Azaarenes: sources, distribution, environmental impact and health effects. In: Saxena J, Fisher F (eds) Hazard Assessment of Chemicals, Vol. 1. Academic Press, New York, pp 421–440.

Schultz TW, Bearden AP (1998) Structure-toxicity relationships for selected naphthoquinones to *Tetrahymena pyriformis*. Bull Environ Contam Toxicol 61:405–410.

Schultz TW, Cajina-Quezada M, Dumont JN (1980) Structure-toxicity relationships of selected nitrogenous heterocyclic compounds. Arch Environ Contam Toxicol 9:591–598.

Seixas GM, Andon BM, Hollingshead PG, Thilly WG (1982) The aza-arenes as mutagens for *Salmonella typhimurium*. Mutat Res 102:201–212.

Shukla OP (1986) Microbial transformation of quinoline by a *Pseudomonas* sp. Appl Environ Microbiol 51:1332–1342.

Siddiqi MA, Ye D, Elmarakby SA, Kumar S, Sikka HC (1994) Microbial metabolism of polycyclic aromatic hydrocarbons (PAH) and aza-PAH. In: Cooke M (ed) Polynuclear Aromatic Hydrocarbons. Battelle Press, Columbus, OH, pp 115–122.

Siim BG, Hicks KO, Pullen SM, van Zijl PL, Denny WA, Wilson WR (2000) Comparison of aromatic and tertiary amine N-oxides of acridine DNA intercalators as bioreductive drugs: cytotoxicity, DNA binding, cellular uptake, and metabolism. Biochem Pharmacol 60:969–978.

Sinks GD, Schultz TW, Hunter RS (1997) UV-B induced toxicity of PAHs: effects of substituents and heteroatom substitution. Bull Environ Contam Toxicol 59:1–8.

Slooff W, Matthijsen AJCM, Montizaan GK, Ros JPM, Van den Berg R, Eerens HC, Goewie CE, Janus JA, Kramers PGN, Van de Meent D, Posthumus R, Schokkin GJH, Wegman RCC, Vaessen HAMG, Wammes JI, Bral EAMA (1989) Integrated criteria document PAHs. RIVM 758474011. National Institute of Public Health and Environmental Protection (RIVM), Bilthoven, The Netherlands.

Southworth GR, Beachamp JJ, Schmieder PK (1978) Bioaccumulation potential and acute toxicity of synthetic fuels effluents in freshwater biota: azaarenes. Environ Sci Technol 12:1062–1066.

Steward AR, Kumar S, Sikka HC (1987) Metabolism of dibenz[a,h]acridine by rat liver microsomes. Carcinogenesis 8:1043–1050.

Sutherland JB, Evans FE, Freeman JP, Williams AJ, Deck J, Cerniglia CE (1994a) Identification of metabolites produced from acridine by *Cunninghamella elegans*. Mycologia 86:117–120.

Sutherland JB, Freeman JP, Williams AJ, Cerniglia CE (1994b) N-oxidation of quinoline and isoquinoline by *Cunninghamella elegans*. Exp Mycol 18:271–274.

Sutherland JB, Freeman JP, Williams AJ (1998) Biotransformation of isoquinoline, phenanthridine, phthalazine, quinazoline, and quinoxaline by *Streptomyces viridosporus*. Appl Microbiol Biotechnol 49:445–449.

Sutton SD, Pfaller SL, Shann JR, Warshawsky D, Kinkle BK, Vestal JR (1996) Aerobic biodegradation of 4-methylquinoline by a soil bacterium. Appl Environ Microbiol 62: 2910–2914.

Swartz RC, Schults DW, Ozretich RJ, Lamberson JO, Cole FA, De Witt TH, Redmond MS, Ferraro SP (1995) ΣPAH: a model to predict the toxicity of polynuclear aromatic hydrocarbon mixtures in field-collected sediments. Environ Toxicol Chem 14:1977–1987.

Tada M, Takahashi K, Kawazoe Y, Ito N (1980) Binding of quinoline to nucleic acid in a subcellular microsomal system. Chem-Biol Interact 29:257–266.

Tada M, Takahashi K, Kawazoe Y (1982) Metabolites of quinoline, a hepatocarcinogen, in a subcellular microsomal system. Chem Pharm Bull 30:3834–3837.

Tafazoli M, Baeten A, Geerlings P, Kirsch-Volders M (1998) In vitro mutagenicity and genotoxicity study of a number of short-chain chlorinated hydrocarbons using the micronucleus test and the alkaline single cell gel electrophoresis technique (Comet assay) in human lymphocytes: a structure-activity relationship (QSAR) analysis of the genotoxic and cytotoxic potential. Mutagenesis 13:115–126.

Talaska G, Roh J, Schamer M, Reilman R, Xue W, Warshawsky D (1995) ^{32}P-postlabelling analysis of dibenz[a,j]acridine DNA adducts in mice: identification of proximate metabolites. Chem-Biol Interact 95:161–174.

Van Genderen J, Noij THM, Van Leerdam JA (1994) Inventory and toxicological evaluation of organic micropollutants. RIWA report. Association of Rhine and Meuse Water Supply Companies (RIWA), Amsterdam, The Netherlands.

Van Leeuwen CJ, Van der Zandt PTJ, Aldenberg T, Verhaar HJM, Hermens JLM (1992) Application of QSARs, extrapolation and equilibrium partitioning in aquatic effects assessment. I. Narcotic industrial pollutants. Environ Toxicol Chem 11:267–282.

Van Vlaardingen PLA, Steinhoff WJ, De Voogt P, Admiraal WA (1996) Property-toxicity relationships of azaarenes to the green alga *Scenedesmus acuminatus*. Environ Toxicol Chem 15:2035–2045.

Veith GD, Mekenyan OG, Ankley GT, Call DJ (1995) A QSAR analysis of substituent effects on the photoinduced acute toxicity of PAHs. Chemosphere 30:2129–2142.

Wakeham SG (1979) Azaarenes in recent lake sediments. Environ Sci Technol 13:1118–1123.

Walton BT, Ho C-H, Ma CY, O'Neill EG, Kao GL (1983) Benzoquinolinediones: activity as insect teratogens. Science 222:422–423.

Wan L, Xue WL, Schneider J, Reilman R, Radike M, Warshawsky D (1992) Comparative metabolism of 7H-dibenzo[c,g]carbazole and dibenz[a,j,]acridine by mouse and rat liver microsomes. Chem-Biol Interact 81:131–147.

Warshawsky D (1992) Environmental sources, carcinogenicity, mutagenicity, metabolism and DNA binding of nitrogen and sulfur heterocyclic aromatics. J Environ Sci Health C 10:1–71.

Warshawsky D, Hollingsworth L, Reilman R, Stong D (1985) The metabolism of dibenz-[a,j]acridine in the isolated perfused lung. Cancer Lett 28:317–326.

Warshawsky D, Barkley W, Miller ML, LaDow K, Andringa A (1994) Carcinogenicity of 7H-dibenzo[c,g]carbazole, dibenz[a,j]acridine and benzo[a]pyrene in mouse skin and liver following topical application. Toxicology 93:135–149.

Warshawsky D, Cody T, Radike M, Reilman R, Schumann B, LaDow K, Schneider J (1995) Biotransformation of benzo[a]pyrene and other polycyclic aromatic hydrocarbons and heterocyclic analogs by several green algae and other algal species under gold and white light. Chem-Biol Interact 97:131–148.

Warshawsky D, Talaska G, Xue W, Schneider J (1996) Comparative carcinogenicity, metabolism, mutagenicity, and DNA binding of 7H-dibenzo[c,g]carbazole and dibenz[a,j]acridine. Crit Rev Toxicol 26:213–249.

Wegener JWM, Klamer JC, Niebeek G (1986) Voorspelling van de toxiciteit van heterocyclische aromatische verbindingen voor aquatische organismen. IVM Report R-86/13. Institute for Environmental Issues (IVM), Amsterdam, The Netherlands.

Wernersson A-S, Dave G (1997) Phototoxicity identification by solid phase extraction and photoinduced toxicity to *Daphnia magna*. Arch Environ Contam Toxicol 32:268–273.

Wiegman S, Van Beusekom SAM, Van Vlaardingen PLA, Bleeker EAJ, Kraak MHS, Admiraal W, Krop H, De Voogt P, Vriezekolk G, Vonck W, Klamer HJC, Pastor D, Evers EHG, Peijnenburg WJGM (1998) Photobiological transformation of azaarenes in the water column. Towards a new approach for the assessment of toxic and genotoxic hazards of azaarenes. BEON 96 M11. Beleidsgericht Ecologische Onderzoek van de Noordzee/Waddenzee (BEON), Amsterdam, The Netherlands.

Wiegman S, Van Vlaardingen PLA, Peijnenburg WJGM, Van Beusekom SAM, Kraak MHS, Admiraal W (1999) Photokinetics of azaarenes and toxicity of phototransformation products to the marine diatom *Phaeodactylum tricornutum*. Environ Sci Technol 33:4256–4262.

Wiegman S, Van Vlaardingen PLA, Bleeker EAJ, De Voogt P, Kraak MHS (2001) Phototoxicity of azaarene isomers to the marine flagellate *Dunaliella tertiolecta*. Environ Toxicol Chem 20:1544–1550.

Wilcke W (2000) Polycyclic aromatic hydrocarbons (PAHs) in soil: a review. J Plant Nutr Soil Sci (Z Pflanzenernaehr Bodenkd) 163:229–248.

Wood AW, Chang RL, Levin W, Ryan DE, Thomas PE, Lehr RE, Kumar S, Schaefer-Ridder M, Engelhart U, Yagi H, Jerina DM, Conney AH (1983) Mutagenicity of diol-epoxides and tetrahydroepoxides of benz[a]acridine and benz[c]acridine in bacteria and in mammalian cells. Cancer Res 43:1656–1662.

Wood AW, Chang RL, Levin W, Kumar S, Shirai N, Jerina DM, Lehr RE, Conney AH (1986) Bacterial and mammalian cell mutagenicity of four optically active bay-region 3,4-diol-1,2-epoxides and other derivatives of the nitrogen heterocycle dibenz[c,h]acridine. Cancer Res 46:2760–2766.

Wood AW, Chang RL, Katz M, Conney AH, Jerina DM, Sikka HC, Levin W, Kumar S (1989) Mutagenicity of dihydrodiols and diol epoxides of dibenz[a,h]acridine in bacterial and mammalian cells. Cancer Res 49:6981–6984.

Yamaguchi K, Ikdea Y, Fueno T (1985) Charge-transfer interactions, exciplex foirmations and ionic dissociations in singlet oxygen reactions. Tetrahedron 41:2099–2107.

Zepp RG (1978) Quantum yields for reaction of pollutants in dilute aqueous solution. Environ Sci Technol 12:327–329.

Zepp RG, Schlotzhauer PF (1979) Photoreactivity of selected aromatic hydrocarbons in water. In: Jones PW, Leber P (eds) Polynuclear Aromatic Hydrocarbons. Ann Arbor Science, Ann Arbor, MI, pp 141–158.

Manuscript received January 9; accepted January 23, 2001.

Enantiomeric Enrichment of Chiral Pesticides in the Environment

Wim J.M. Hegeman and Remi W.P.M. Laane

Contents

I. Introduction	86
II. Methodology and Data Selection	88
A. Enantiomeric Ratios	88
B. Enantiomer Fractions	88
C. Measurements	89
D. Database	89
III. Selected Chiral Compounds	93
A. α-HCH	93
B. Mecoprop	94
C. Chlordane Compounds	95
IV. Results	95
A. α-HCH	95
B. Mecoprop	96
C. *cis*-Chlordane	97
D. *trans*-Chlordane	98
E. Heptachlor *exo*-Epoxide	99
F. Oxychlordane	100
V. Discussion	100
A. Deviations from the Racemic Mixture	100
B. Constant Enantiomer Fraction	101
C. Model Scheme	102
D. Stereochemical Recognition of Chlordane Compounds	103
E. Shielding from the Racemate	104
F. Chiral Enrichment Processes	105
G. Environmental Regulations	106
Summary	106
Acknowledgments	107
References	107

Communicating Editor: Pim de Voogt.

W.J.M. Hegeman (✉)
Water Research Foundation, Johan Wagenaarlaan 4, 2132 KE Hoofddorp, The Netherlands

R.W.P.M. Laane
Department of Environmental and Toxicological Chemistry, University of Amsterdam, Nieuwe Achtergracht 166, 1018 WV Amsterdam, The Netherlands and Ministry of Transport, Public Works and Water Management, National Institute for Coastal and Marine Management, P.O. Box 20907, 2500 EX The Hague, The Netherlands

I. Introduction

Many agrochemicals consist of chiral compounds and about 25% of all agrochemicals used worldwide are chiral compounds (Williams 1996). In the Netherlands, approximately 23% of the total number of 265 permitted agrochemicals consist of a mixture of chiral compounds (CTB 1999). Only a small fraction of all agrochemicals are manufactured and used in the form of a pure enantiomeric compound. For example, in Switzerland and the Netherlands, the only type of chiral phenoxy herbicides permitted is the type containing only the active enantiomer (Williams 1996). For compounds that possess more than one chiral center, more than half of the compound may enter the environment as enantiomer ballast (Kohler et al. 1997; Eckhardt et al. 1992; Moser et al. 1982). There are advantages to be gained from using pure enantiomers rather than the racemate. Enantiomeric agrochemicals, in which one enantiomer is significantly more active than the other, are likely to cost less to produce, can be used in smaller quantities, and cause less environmental damage (Spangler et al. 1999; Renner 2000). Asymmetrical synthesis of pure enantiomers can be profitable because the patents of the racemic compounds expire in due course and because pure enantiomers are authorized exclusively in crop protection (Blaser and Spindler 1997; Williams 1996). However, most of the chiral agrochemicals found in the environment originate from symmetrical synthesis and are applied as such in the field.

Organisms produce or degrade chiral compounds by stereospecific enzymatic processes (Wiberg et al. 1998a). Therefore, application of chiral mixtures may result in a change in the concentration of the enantiomers that deviates from the racemic composition originally applied (Kohler et al. 1997; Faller et al. 1991a). As a result, the enantiomeric ratio (ER), defined as the ratio of the concentrations of the (+)-enantiomer and the (−)-enantiomer, will deviate from the racemic composition (ER = 1) (Vetter and Schurig 1997).

Enantiomer enrichment processes for chiral drugs in humans and laboratory animals have been studied extensively (Simonyi 1984; Brossi 1994; Caldwell 1996). Deviations from the racemic mixture have been found in plasma protein binding (Bertilsson et al. 1991; Hashimoto et al. 1992; Nordin and Bertilsson 1995; Rochat et al. 1995; Baudry et al. 1997; Sidhu et al. 1997; Yu et al. 1997), brain tissue (Hashimoto et al. 1992; Fuller and Snoddy 1993; Aoyama et al. 1994; Aspeslet et al. 1994; Baldessarini et al. 1994; Nybäck et al. 1994; Scanley et al. 1994; Baudry et al. 1997; Sam et al. 1997; Yu et al. 1997), brain fluid (Smith and Peterson 1982; Bertilsson et al. 1991; LaManna et al. 1993; Nordin and Bertilsson 1995; Sam et al. 1997), liver tissue (Püttman et al. 1989; Nordin and Bertilsson 1995; Rochat et al. 1997), and kidney tissue/urine (McErlane et al. 1990; Aspeslet et al. 1994; Prien et al. 1997). However, despite considerable pharmacological research, the processes such as stereoselective active transportation through membranes, stereoselective degradation, and stereoselective binding are not completely understood on the molecular scale. In general, it is assumed that active sites are able to select monomers of matching chirality from a racemic mixture, thus giving rise to the formation of diastereomeric complexes between the active sites of enzymes and the chiral compounds (Hühnerfuss 2000). Diastereomeric complexes differ in their physical properties (Ternay 1979).

Abiotic enantioselective transformations can also change the ER. However, for these enantioselective processes a chiral catalyst or an excess of one enantiomer from a reactive chiral compound is required (March 1985). However, no such process condition has been reported in enantiomeric environmental studies (Zipper et al. 1998a). Enantiomerization reactions, which can also lead to changes in the ER, are biologically mediated (Buser and Müller 1998; Reist et al. 1995).

The ER can be used as a tracer tool in environmental studies (Bidleman 1998; Bidleman and Falconer 1999a,b). A distinction can be made between "old" and "new" sources of a chemical, and the fate of a compound can be traced in water and air (Bidleman et al. 1998a; Jantunen et al. 1998; Falconer et al. 1998; Ridal et al. 1997). For example, enantiomer profiles of nonracemic pesticide residues in soil and water are preserved on volatilization; the "old source" signature can be distinguished from a freshly applied racemic pesticide. Even an abnormal source of chiral herbicides, such as roof sealing as a source of mecoprop,[1] can be identified by measuring ERs (Bucheli et al. 1998a,b).

In general, hydrophobic persistent organic compounds accumulate from a lower trophic level to a higher trophic level. In a food web, the concentration increases to the highest trophic level (de Voogt 1996). Wiberg et al. (1998a, 2000) showed that concentrations of the enantiomers of α-HCH (hexachlorocyclohexane)[1] and chlordane compounds became biomagnified in the polar bear food chain. Additionally, ERs increased from total cod (ER $\approx$ 1) to blubber and liver samples of ringed seals to liver samples of polar bear (ER = 2.3). Multivariate statistical methods have been used to investigate the relationships between ERs, chemical residue concentration, and biological data (Wiberg et al. 2000). ERs of α-HCH and chlordane compounds were the most important variables for the sample groupings and for the class separation of male/female seals and fat/liver tissues (Wiberg et al. 2000). Therefore, ER measurements are likely to be a valuable distinctive tool for food chain analysis. Tanabe et al. (1996) showed that the ERs of α-HCH increased from ER = 0.8–1 for water, air, and lower trophic levels to ER = 1.6–2.8 for higher trophic levels such as pinnipeds and blubber of cetaceans. Iwata et al. (1998) studied the biological and ecological factors of enantioselective accumulation of α-HCH. They found that the ERs in higher trophic animals were influenced by species-specific metabolism and transport processes in the body as well as by biological factors, whereas the ERs were also changed by ecological factors such as feeding habits. Deviations in ER as a result of sexual maturity, aging, and breeding activities were not significant (Iwata et al. 1998).

---

[1]For precise chemical nomenclature and Chemical Abstract Numbers (CAS), see the Appendix.

Changes in ER caused by differences in habitat and food sources were also found by Kallenborn et al. (1998). These results are in contrast to those found by Wiberg et al. (1998b), who showed that the sampling site (habitat) played a minor role in ER changes. In general, changes in ER are larger in seals than in herrings. According to Vetter and Muraya (2000), this is true because in higher organisms pollutants are subjected to an increased specialization of enzyme systems. Also, the health status of the animal may be a factor in the enantiomer-specific metabolism of chiral pesticides (Wiberg et al. 1998b). Wiberg et al. (1998b) showed that species-specific differences, e.g., inverse ERs of *cis*-chlordane[1] and *trans*-chlordane[1] in ringed seal versus harbor and grey seal, are important factors controlling ERs in Baltic seals. Their conclusions agree with results reported in the work of Iwata et al. (1998). Karlsson et al. (2000) showed that ERs of *cis*-chlordane and *trans*-chlordane were gender specific in cod. The female cod preferentially accumulated (−)-*cis*-chlordane and (−)-*trans*-chlordane.

ERs found in environmental compartments change in time and place (Zipper et al. 1998a; Jantunen and Bidleman 1998), but a theory explaining these changes has not yet been developed. The purpose of this review is to develop a model to predict enantiomeric enrichment in the environment for persistent pesticides. The model will be based on published ERs for six chiral compounds in air, water, soil, and biotic compartments.

II. Methodology and Data Selection
A. Enantiomeric Ratios

An enantiomeric ratio (ER) is defined as the (+)-enantiomer concentration of a chiral compound divided by its (−)-enantiomer concentration (Vetter and Schurig 1997):

$$ER = \frac{(+)\text{-}enantiomer}{(-)\text{-}enantiomer} \qquad (1)$$

The (+)-sign indicates that the enantiomer rotates a plane of polarized light to the right, clockwise (March 1985). The isomer with the (−)-sign rotates the light to the left, counterclockwise. Chiroptical detection is not always included in chiral chromatographic analysis. Also, pure enantiomers or enantiomeric-enriched reference materials may not be available; therefore, the enantiomer elution sequence cannot be achieved. In such cases, the ER is defined as the concentration of the first eluting enantiomer divided by the second enantiomer, under precisely defined chromatographic conditions (Vetter and Schurig 1997). Nowadays, pure enantiomer or enantiomer-enriched standards are often available and the elution sequence can be easily determined (Müller and Buser 1994). In this study, we use the ERs as presented in the literature.

B. Enantiomer Fractions

The enantiomeric ratio data were transformed into enantiomer fractions (EFs) as a standard descriptor (Harner et al. 2000). The EF can be calculated from ER by the formula:

$$EF = \frac{ER}{ER + 1} \tag{2}$$

This descriptor provides a more meaningful representation of graphical data than the ER and is more easily employed in mathematical fate expressions (Harner et al. 2000). On the basis of an earlier draft version of this article, we found that the graphical representation of EFs is superior to the plotting of ERs and to the plotting of ERs in combination with inverse ERs (1/ER). EFs can always be plotted, even for ERs that reach infinity; the EF can only range from 0 to 1.0, with $EF = 0.5$ representing the racemic mixture.

C. Measurements

Measurements of the ERs of chiral organochlorine compounds in environmental samples started around 1990 (König et al. 1991) when highly selective chiral chromatographic columns were used. Chiral column material based on cyclodextrines yielded good separation of enantiomers (Vetter and Schurig 1997). Nevertheless, separation of enantiomers is not always successful. Chiral compounds with C-asymmetry are often easier to separate than compounds with axial asymmetry, known as atropisomers (Buser and Müller 1995a). Chiral gas chromatography was generally used to separate chiral organochlorine compounds in very low quantities, but electrophoresis (El Rassi 1997; Garrison et al. 1996; Penmetsa et al. 1997), micellar electrokinetic chromatography (Schmitt et al. 1997), and HPLC (Sevcik et al. 1997) have also been used.

D. Database

Data were selected from publications in which enantiomers were determined in different environmental compartments: water (dissolved phase and suspended particulates), air, soil, and biota. Only field data were used; no results from laboratory studies using field material were incorporated (Tett et al. 1994, 1997; Buser and Müller 1997, 1998; Zipper et al. 1998a,b; Zipper 1998). The references for the ERs are grouped by the chiral compounds in Table 1.

Six xenobiotics—α-HCH, mecoprop, *cis*-chlordane (CC), *trans*-chlordane (TC), oxychlordane (OXY), and heptachlor *exo*-epoxide (HEPX)—were selected from Table 1. These compounds were measured in several environmental compartments. The absolute configuration of the (+)- and (−)-enantiomers are shown in Fig. 1.

Table 2 lists the numbers of samples for the compartments water, air, soil, and biota. The ERs of the same origin (location, organism, organ) were grouped and an average ER was calculated. Every group or compartment is given a compartment number, n. For α-HCH, the total of 618 separate ERs could be grouped in 99 different compartments. Statistical calculations were performed in the database but were not used in the figures because they did not improve the distinctness. EFs were calculated according to Eq. 2 (Harner et al. 2000). The complete data set contains ER, organism name, the organs on which analy-

Table 1. List of chiral compounds with references.

Compound	Reference
α-HCH (hexachlorocyclohexane)	Faller et al. (1991a,b)
	Kallenborn et al. (1991, 1998)
	Müller et al. (1992)
	Hühnerfuss et al. (1992a,b, 1993)
	Mössner et al. (1992)
	Pfaffenberger et al. (1992, 1994a)
	Möller et al. (1993)
	Hummert et al. (1995)
	Falconer et al. (1995a,b, 1997)
	Jantunen and Bidleman (1996, 1997, 1998)
	Finizio et al. (1998)
	Klobes et al. (1998a)
	Oehme et al. (1995)
	Tanabe et al. (1996)
	Ridal et al. (1997)
	Aigner et al. (1998)
	Iwata et al. (1998)
	Wiberg et al. (1998a,b)
	Jantunen et al. (1998)
	Harner et al. (1999)
Phenoxypropanoic acids	Buser and Müller (1998)
(e.g. mecoprop and dichlorprop)	Buser et al. (1998)
	Bucheli et al. (1998a,b)
	Zipper et al. (1998a)
Chlordane compounds:	Buser et al. (1992)
cis-Chlordane (CC)	Büser and Müller (1993)
	Falconer et al. (1997)
	Müller et al. (1997)
	Wiberg et al. (1997, 1998a)
	Jantunen and Bidleman (1998)
	Aigner et al. (1998)
	Ulrich and Hites (1998)
trans-Chlordane (TC)	Aigner et al. (1998)
	Büser and Müller (1993)
	Buser et al. (1992)
	Falconer et al. (1997)
	Jantunen and Bidleman (1998)
	Müller et al. (1997)
	Ulrich and Hites (1998)
	Wiberg et al. (1997, 1998a)

Table 1. Continued.

Compound	Reference
Heptachlor *exo*-epoxide[1] (HEPX)	Hühnerfuss et al. (1993)
	Büser and Müller (1993)
	König et al. (1994)
	Müller et al. (1997)
	Finizio et al. (1998)
	Falconer et al. (1997)
	Wiberg et al. (1997, 1998a)
	Aigner et al. (1998)
	Bidleman et al. (1998b)
	Ulrich and Hites (1998)
	Jantunen and Bidleman (1998)
	Pfaffenberger et al. (1994a)
Oxychlordane[1] (OXY)	Hühnerfuss et al. (1993)
	König et al. (1994)
	Müller et al. (1997)
	Aigner et al. (1998)
	Klobes et al. (1998a)
	Pfaffenberger et al. (1994a)
	Vetter et al. (1997a)
	Wiberg et al. (1998a)
β- and γ- pentachlorocyclohexene atropisomers of PCBs	Hühnerfuss et al. (1992b, 1993)
	Vetter et al. (1997a)
	Glausch et al. (1995)
	Hühnerfuss et al. (1995)
	Blanch et al. (1996)
	Haglund and Wiberg (1996)
	Reich et al. (1999)
	Klobes et al. (1998a)
Atropisomer of methylsulfonyl-PCBs	Ellerichmann et al. (1998)
	Bergman et al. (1998)
	Wiberg et al. (1998c)
Toxaphene/polychlorinated bornanes	Kallenborn et al. (1994)
	Vetter et al. (1997a,b, 1998, 1999)
	Klobes et al. (1998b)
2,4'-DDD and 2,4'-DDT	Buser and Müller (1995a)
	Falconer et al. (1997)
	Finizio et al. (1998)
	Aigner et al. (1998)
Bromocyclen	Pfaffenberger et al. (1994b)

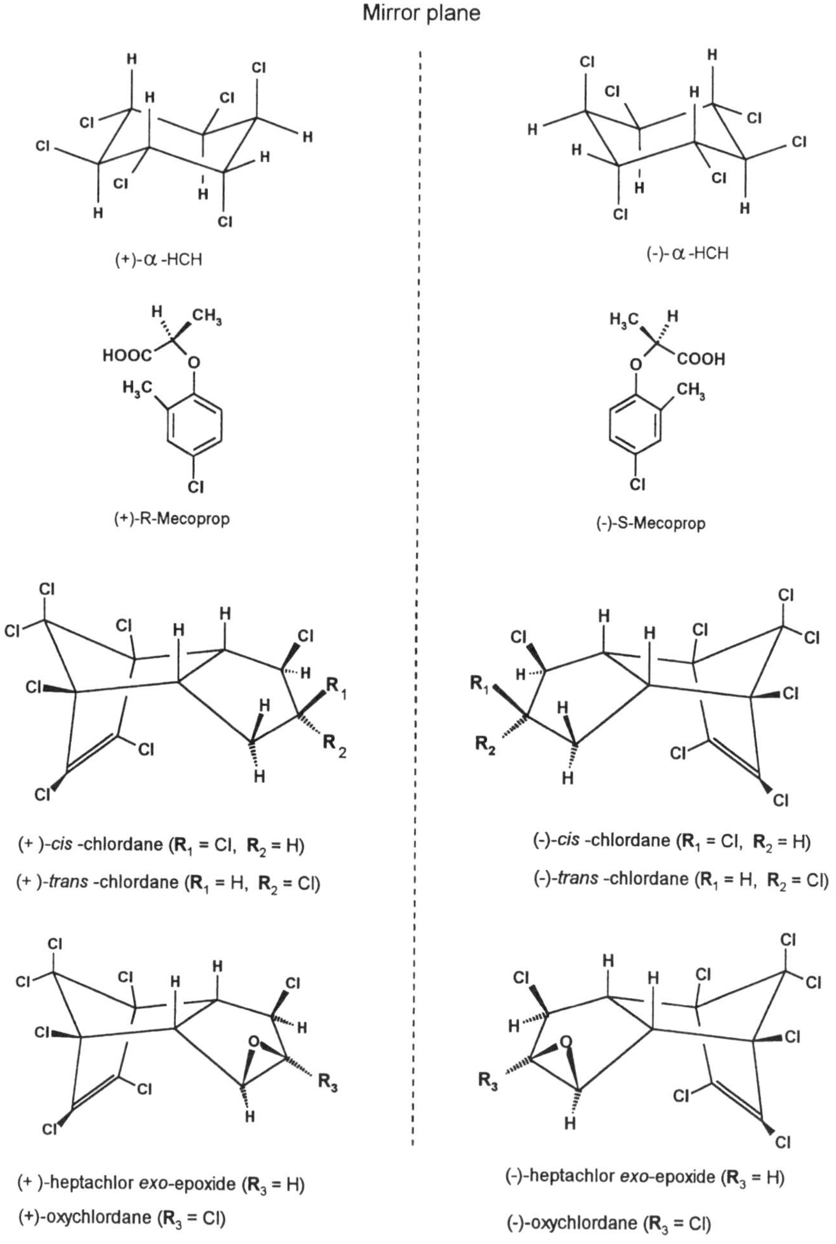

Fig. 1. Structures of the enantiomers of α-hexachlorocyclohexane (α-HCH), mecoprop, *cis*-chlordane, *trans*-chlordane, heptachlor *exo*-epoxide, and oxychlordane. The (+)-enantiomer cannot be superimposed upon its mirror image, the (−)-enantiomer. Note: α-HCH has *aaeeee*-conformation. Absolute configuration is according to Buser and Müller (1995b) (α-HCH); Zipper et al. (1998b) (mecoprop), and Miyazaki et al. (1980) (chlordane compounds).

Table 2. Number of samples in different compartments for which enantiomeric ratios (ER) were determined.

Compound	Total	Water	Air	Soil	Biota
α-HCH	618	269 (12)[a],(7)[b]	89 (9)[c]	10	250
Mecoprop	92	92[d]	—	—	—
cis-Chlordane	86	11	16	31	28
trans-Chlordane	96	14	17	32	33
Heptachlor exo-epoxide	105	12	17	26	50
Oxychlordane	99	—	—	17	82

—, no values found.
[a]Suspended particulates are assigned to the water compartment.
[b]Snow is assigned to water compartment.
[c]Rain is assigned to the air compartment.
[d]Groundwater samples belong to the water compartment.

sis was made, sampling area, date (when available), and the references. This data set is available on the Internet at **HTTP://www.waterresearch.nl/ERDATA.HTM** or from the author on request.

III. Selected Chiral Compounds
A. α-HCH

Technical hexachlorocyclohexane (HCH) used to be an important insecticide in agriculture and in forestry, and served as a wood preservative (Slooff and Matthijsen 1988). The technical mixture of HCH, which is produced by chlorination of benzene under UV light, consists of 60%–70% α-HCH, 5%–12% β-HCH, 10%–15% γ-HCH, and 6%–10% δ-HCH and smaller amounts of other isomers and congeners. The cumulative world emission of technical HCH between 1947 and 1997 amounted to 6800 kton (Wania and Mackay 1999). The use of technical HCH was discontinued in the United States in 1978 (Falconer et al. 1995b) and other industrialized countries but continues in Third World countries (Buser and Müller 1995a). Nine stereoisomers of 1,2,3,4,5,6-hexachlorocyclohexane, seven mesoforms and a dl-pair, exist in theory (March 1985). Eight unique isomers can be derived: α-, β-, γ-, δ-, ε-, η-, θ-, and ι-HCH (Deo et al. 1994). Only α-HCH consists of a pair of enantiomers. γ-HCH or lindane is an insecticide (Slooff and Matthijsen 1988). The chair conformer of HCH is stable at room temperature. The six chlorosubstituents are either in axial (a) or in equatorial (e) positions. The chair conformer can be involved in chair–chair interconversion, also called cyclohexane ring inversion (Ternay 1979). In this process, all the bonds that are axial become equatorial whereas those that are equatorial become axial. Therefore, α-HCH with the *aaeeee*-conformation is equal to α-HCH with the *eeaaaa*-conformation. The equatorial position is the more favorable one for a relatively large substituent such as chlorine. The molecule will acquire the conformation in which most chlorine atoms are situated in an equatorial position. For α-HCH, *aaeeee* is the preferred conformation (Fig. 1).

Distribution and Metabolism. Relatively high concentrations of HCHs (sum of α- and γ-HCH) in air samples were found in the Northern hemisphere in tropical Asia (Bay of Bengal and Arabian Sea: 690–32,000 pg/m^3; Iwata et al. 1993). Water samples showed a considerable increase in HCH concentration above 40° N latitude in the North Pacific (Northern North Pacific: 240–1600 pg/L; Iwata et al. 1993). The global distribution of HCH isomers is controlled by their physicochemical properties and by meteorological conditions (Iwata et al. 1993). α-HCH is the most stable isomer on photolysis in the presence of iron salts (Malaiyandi and Shah 1984), which explains the persistence of α-HCH in the environment. Deo et al. (1994) concluded that any stereoisomer of HCH added to the environment is vulnerable to interconversion and degradation, processes that continue until an equilibrium is reached between the different isomers. The toxicity and the fate of HCH isomers in humans, wildlife, plant, soil, water, and atmosphere were reviewed by Willett et al. (1998). They concluded that the reasons for differences in the enantiomeric enrichment of α-HCH in environmental and biological samples are still largely unclear.

B. Mecoprop

Mecoprop belongs to the phenoxyalkanoic acids, or, simply, the phenoxy herbicide group. It is the most widely used herbicide for broadleaf weed control in cereal crops throughout the world (Worthing and Hance 1991). In the U.S., 23,000 kt of mecoprop and dichlorprop are used annually (Schneiderheinze et al. 1999). Of the total amount of mecoprop produced in the European Union (5000 t/yr), only 5% is applied in the form of the pure, active enantiomer (+)-(R)-mecoprop (Zipper 1998). The phenoxyalkanoic herbicides are highly water soluble acids (pK_a = 3.11) and have a low tendency to accumulate in organic matter. At pH 7, they are not sorbed to most soil types (Zipper et al. 1998a).

Distribution. Residues of phenoxyalkanoic herbicides are often found in subsurface and groundwater samples (Felding 1995; Fielding et al. 1991; Gintautas et al. 1992). Because of the hydrophilic character of Mecoprop, all ER measurements from field studies were in the dissolved phase (see Table 2). In European countries (Denmark, Germany, Great Britain, The Netherlands, Italy, Sweden), residues of (R,S)-mecoprop were found in drinking water in concentrations higher than the maximum allowable concentration for an individual pesticide, i.e., 100 ng/L (Zipper 1998).

C. Chlordane Compounds

Technical-grade chlordane was used extensively as a pesticide in the U.S. from 1948 to 1988 (ATSDR 1994). The cumulative world production is 70,000 tons (Christen 1999). Technical chlordane consists of more than 100 different compounds that are structurally related (Vetter and Schurig 1997). *cis*-Chlordane, *trans*-chlordane, and heptachlor are the major compounds, accounting for 19%, 24%, and 22% of technical chlordane, respectively (Verschueren 1996). *cis*- and *trans*-Chlordane (CC and TC) are the insecticidally relevant parts of the formulations (Vetter and Schurig 1997). Heptachlor *exo*-epoxide (HEPX) was tested as an insecticide, (+)-HEPX exhibiting stronger insecticidal activity than (−)-HEPX (Miyazaki et al. 1978). Oxychordane (OXY) is the common metabolite of *cis*- and *trans*-chlordane in biota (Vetter and Schurig 1997). Oxychlordane can also be formed from octachlors and nonachlors (Aigner et al. 1998; Müller and Buser 1994). Heptachlor *exo*-epoxide (HEPX) is the major metabolite of heptachlor. The epoxidation of heptachlor occurs stereoselectively with conversion of the absolute configuration (Müller and Buser 1994).

Distribution. Chlordane compounds (sum of TC, CC, and *trans*-nonachlor) have relatively high Henry's law constant values. Consequently, atmospheric transport is an important migration route (Iwata et al. 1993). Chlordane compounds can be deposited in the Arctic and adjacent water bodies. Air concentrations of 8.1–160 pg/m^3 were found in the South China Sea and 5.5–55 pg/m^3 in the northern North Pacific (Iwata et al. 1993). Aqueous concentrations of chlordane compounds (sum of TC, CC, and *trans*-nonachlor) amounted to 3.9–22 pg/L in the East China Sea and 4.3–17 pg/L in the northern North Pacific (Iwata et al. 1993).

IV. Results
A. α-HCH

According to the literature, α-HCH is the compound with the highest number of measurements (Table 2). The average enantiomer fractions (EFs) in a group n of the same origin are plotted in ascending order in Fig. 2. The racemic composition has an EF of 0.5 (line). Compartments with an EF higher than 0.5

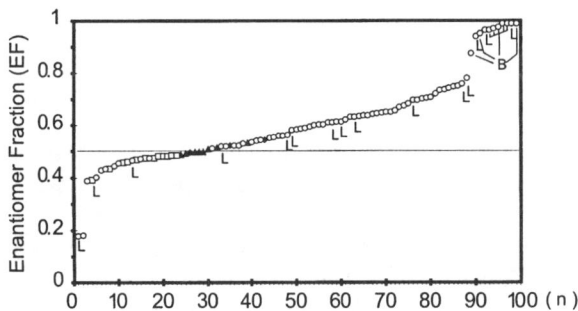

Fig. 2. Average enantiomer fraction (EF) of α-HCH in soil (♦), air (▲), water (□), and biotic compartments (○) in ascending order by compartment number (n). L, liver tissue; B, brain tissue.

have higher (+)-enantiomer concentrations, whereas EFs below the line EF = 0.5 have higher (−)-enantiomer concentrations (Fig. 2). Equal distances above and below the line EF = 0.5 indicate equal but opposite stereoselective behavior. There are two main groups, the abiotic compartments consisting of soil, air, and water and the biotic compartments consisting of all data from different organisms and organs. Extremely high EFs for the biotic compartments are shown in the upper right section of Fig. 2 whereas low EFs are shown in the lower left section. The EF of α-HCH in Arctic particulate material (Fig. 2, $n = 14$) was assigned to the water compartment because it interacts strongly with water. On the basis of studies of mecoprop behavior (Zipper et al. 1998a), we assume that sorption processes for chiral compounds are nonstereoselective.

In general, the abiotic compartments showed average EFs close to 0.5 (from 0.39 to 0.55; Fig. 2), whereas average EFs in biotic compartments showed a much higher deviation, from EF = 0.5 (0.18–0.99; Fig. 2). In biota, the enrichment in either (+)-α-HCH or (−)-α-HCH is often higher than in soil, air, and water. Most soil compartments (Fig. 2) showed a higher (+)-α-HCH concentration than their enantiomer because the (−)-α-HCH degrades faster than (+)-α-HCH (Finizio et al. 1998; Falconer et al. 1997). Nonenantiomer-selective photochemical reactions from γ-HCH are additional sources of α-HCH (Müller et al. 1992). Enantiomer enrichment of α-HCH from γ-HCH is mediated microbially (Faller et al. 1991a). In most cases, the air compartment (Fig. 2) shows EFs close to the racemic mixture of α-HCH. Only Arctic samples ($n = 30, 32$) and air samples taken above agricultural areas ($n = 39$) yielded slightly higher EFs. The aqueous compartment shows values close to EF = 0.5 or just below EF = 0.5, indicating that (+)-α-HCH degrades faster than (−)-α-HCH. The biotic compartments show EFs that are generally much higher than 0.5. Biotic compartments with EF < 0.5 are roe-deer liver ($n = 1$), blubber of hooded seal (Arctic) ($n = 2$), blubber of Antarctic Weddell seals ($n = 3$), Baltic blue mussel ($n = 5$), hooded seal from the North Sea ($n = 6$), fat ($n = 7$), and liver ($n = 9$) of sheep, herring $n = 10$), liver of flounder ($n = 13$), blue mussel from the North Sea ($n = 15$), and cod ($n = 17$). Liver and brain organs, especially in seal and bird samples, show EFs close to 1 ($n = 86–99$; Fig. 2); this indicates strong enrichment of (+)-α-HCH in these compartments.

B. Mecoprop

Average enantiomer fractions (EFs) of mecoprop are plotted in ascending order in Fig. 3. The majority of the studies were performed in Swiss groundwater and surface waters. Mecoprop is one of the few pesticides applied as a pure enantiomer (Williams 1996). In Switzerland, only (+)-*R*-mecoprop is registered for agricultural use (Williams 1996). Therefore, it was expected that EFs close to 1 would be found; in fact, EFs of less than 0.5 were found in Swiss wastewater treatment plants ($n = 1$), Swiss streams ($n = 2, 3$), roof water ($n = 4, 8$), and lakes ($n = 6, 7, 10, 11$) (Buser and Müller 1998; Bucheli et al. 1998a,b). These EFs are the result of selective breakdown of the *R*-enantiomer. EFs of less than

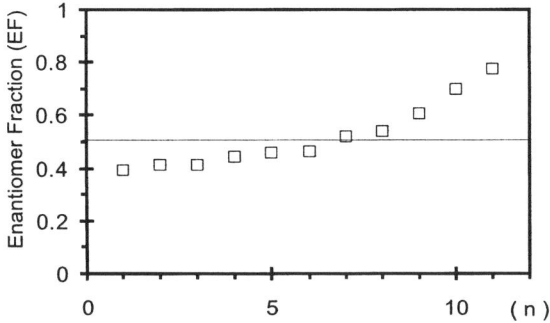

Fig. 3. Average enantiomer fraction of mecoprop in water compartments (□) in ascending order by compartment number (n).

0.5 were also found in the North Sea ($n = 5$) (Buser et al. 1998). The highest EFs ($n = 9$; Fig. 3) were found in groundwater wells at a former chemical waste disposal site at Kölliken, Switzerland (Zipper et al. 1998a). A range of EFs were found here but they depended on the site, the depth of the well, and sampling time. High EFs were often found in deep wells sampled in November, whereas in the same wells in February some EFs were close to 0.5 (Zipper et al. 1998a). High average EFs ($R > S$), up to 0.78, were found in Lakes Sempachersee ($n = 10$) and Hallwilersee ($n = 11$) in 1997 (Buser and Müller 1998). These values are the result of agricultural practices in combination with degradation of the R-enantiomer in soils (Buser and Müller 1998). The lowest EF (0.39) (Fig. 3) was from a wastewater treatment plant near Lake Greifensee ($n = 1$); the S-enantiomer had the highest concentration (Bucheli et al. 1998b). Apparently a bituminous roof sealing membrane impregnated with a Preventol B2 rubber seal protectant against root penetration was responsible for the high concentration of racemic mecoprop detected in municipal wastewater in Grüze, Switzerland (Bucheli et al. 1998a). Preventol B2 is a technical product of polyethylene glycol and releases (R,S)-mecoprop when hydrolyzed (Bucheli et al. 1998a).

C. cis-Chlordane (CC)

Average enantiomer fractions (EFs) of *cis*-chlordane are plotted in ascending order in Fig. 4. Soil samples in agricultural areas showed EFs higher than 0.5 ($n = 6$, 11–16). Air compartments in Southern Norway ($n = 8$), Southern U.S. ($n = 9$), and the Great Lakes ($n = 10$) show EFs of approximately 0.5. The deviation from EF = 0.5 in air compartments is very small (range, 0.50–0.52). The EFs in the surface water from the Arctic (Jantunen and Bidleman 1998) were close to 0.5 ($n = 7$). EFs in biotic compartments of *cis*-chlordane varied from 0.15 ($n = 1$) to 0.72 ($n = 20$). Average EFs in harbor seal blubber ($n = 1$), grey seal liver ($n = 2$), grey seal blubber ($n = 4$), and Baltic salmon ($n = 3$) samples showed relatively higher concentrations of the (−)-enantiomer; average EFs of

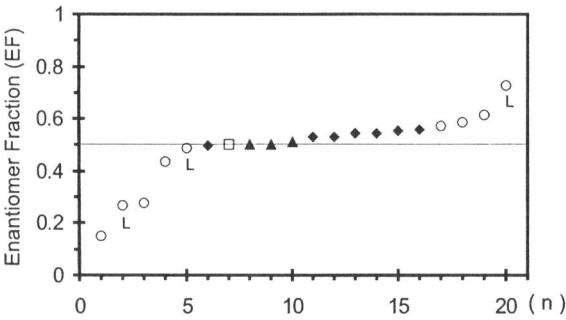

Fig. 4. Average enantiomer fraction of *cis*-chlordane (CC) in soil (◆), air (▲), water (□), and biotic compartments (○) in ascending order by compartment number (*n*). L, liver tissue.

herring ($n = 18$) and ringed seal blubber ($n = 19$) and liver ($n = 20$) of samples showed higher concentrations of the (+)-enantiomer concentration.

D. *trans*-Chlordane (TC)

Average *trans*-chlordane EFs are plotted in ascending order in Fig. 5. The EF in all agricultural soil compartments ($n = 11–15$, 18, 19; Fig. 5) is below the line EF = 0.5, which indicates that an enantioselective process has been in operation. The air compartments showed EFs close to 0.5 ($n = 17$, 20, 21; Fig. 5). The air above the Great Lakes (Ontario, Michigan, Erie, Superior; $n = 17$) showed an average EF of 0.47. Wiberg et al. (1997) proposed three possible sources for *trans*-chlordane in air: termiticides used in households, long-range transport from regions where chlordane is still in use, and, finally, volatilization from agricultural soil or from the Lakes themselves. Air samples from the Southern

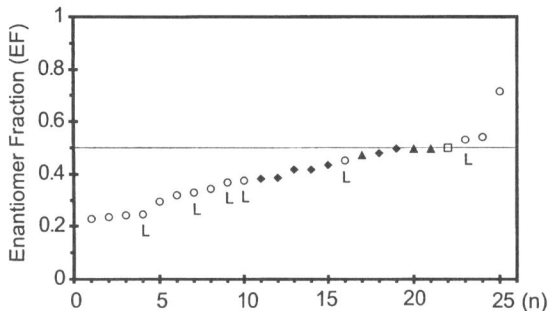

Fig. 5. Average enantiomer fraction of *trans*-chlordane (TC) in soil (◆), air (▲), water (□), and biotic compartments (○) in ascending order by compartment number (*n*). L, liver tissue.

U.S. (Alabama, South Carolina; $n = 21$) were close to 0.5 and showed no enantiomeric enrichment (Wiberg et al. 1997). Indoor air showed a high concentration of chlordane compounds, probably resulting from the use of termiticides (Wiberg et al. 1997). The EFs in air from the Norwegian south coast ($n = 20$) were 0.49 and showed no enantiomeric enrichment. It was concluded that the decline in concentration could only result from abiotic degradation processes of chlordane compounds (Buser and Müller 1993). Surface water in the Arctic showed EFs close to 0.5 (range, 0.49–0.51). In the Chukchi Sea, some values for EF were less than 0.5 and in the Greenland Sea higher EFs were more than 0.5 (Jantunen and Bidleman 1998), indicating reversed enantiospecific processes in different seas. *trans*-Chlordane enantiomers in biotic samples showed the EF to be less than 0.5 in several compartments ($n = 1–10$). Most organs in different seal species ($n = 1–4, 7–10, 16$) from Swedish waters showed enantiomeric enrichment of the (−)-enantiomers (Wiberg et al. 1998b; Buser et al. 1992). In addition, herring ($n = 2, 3, 6$), which is a food source for seals, showed an EF of less than 0.5 (Wiberg et al. 1998a). Salmon caught in the Swedish River Ume ($n = 24$) showed an EF greater than 0.5 in muscle (Buser et al. 1992). Blubber of the harbor seal ($n = 25$) in Swedish waters showed EFs greater than 0.5 (Wiberg et al. 1998a). These EFs showed enantiospecific processes to be the opposite of those found in liver samples of harbor seal ($n = 9$).

E. Heptachlor *exo*-Epoxide (HEPX)

Average heptachlor *exo*-epoxide EFs are depicted in ascending order in Fig. 6. Soil samples from the Fraser Valley, Canada ($n = 8$) showed EFs of 0.52 (Falconer et al. 1997). Samples from the U.S. agricultural areas showed EFs between 0.73 and 0.75 ($n = 17–21$). The average EFs in air compartments were between 0.58 and 0.67 ($n = 9, 10, 13, 14$). The highest EFs were found in air from Lake Ontario (0.65) and Lake Superior (0.67) (Wiberg et al. 1997). The average EF of HEPX in surface water at the North Pole ($n = 12$) was 0.62

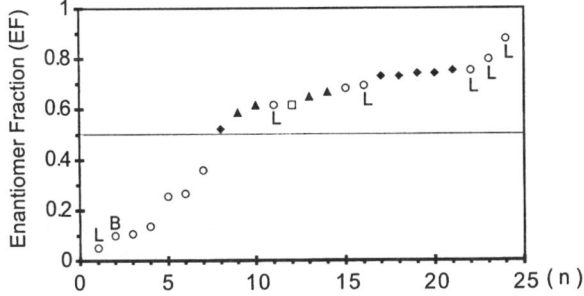

Fig. 6. Average enantiomer fraction of heptachlor *exo*-epoxide (HEPX) in soil (◆), air (▲), water (□), and biotic compartments (○) in ascending order by compartment number (n). L, liver tissue; B, brain tissue.

(Jantunen and Bidleman 1998). The EFs of HEPX in biotic compartments ranged from 0.05 to 0.14 for the organs liver ($n = 1$), brain ($n = 2$), and blubber ($n = 4$) in seals from Iceland (König et al. 1994). Blubber samples of harbor seal ($n = 3$), grey seal ($n = 5$), and ringed seal ($n = 6$) from Swedish waters yielded EFs below 0.5 (Wiberg et al. 1998b). On the other hand, samples of seagull ($n = 15$), hare liver ($n = 22$), rat liver ($n = 23$), and roe-deer liver from Baden-Württemberg and Schleswig-Holstein ($n = 24$) yielded EFs higher than 0.5 (König et al. 1994; Pfaffenberger et al. 1994a; Buser and Müller 1993.

F. Oxychlordane (OXY)

Average EFs of oxychlordane are depicted in ascending order in Fig. 7. EFs for OXY were determined only in soil and in biotic compartments. Average EFs of agricultural soils were higher than 0.5 ($n = 11, 13, 19$), except for Illinois soils ($n = 7$) (Aigner et al. 1998). The lowest EF was found in liver of harbor seal ($n = 1, 2$) and in brain of seal ($n = 3$) (Wiberg et al. 1998b; König et al. 1994). Exceptionally high average EFs of 0.92 and 0.95 were found in roe-deer liver samples from the German states Schleswig-Holstein ($n = 22$) and Baden-Württemberg ($n = 23$), respectively (Pfaffenberger et al. 1994a).

V. Discussion
A. Deviations from the Racemic Mixture

From the literature, we selected the enantiomer ratios (ERs) of six chiral organochlorine pesticides that were measured in air, water, soil, and biota. Enantiomeric processes in a compartment cause the enantiomer fraction (EF) to deviate from the racemic composition (EF = 0.5). For air, water, soils, and organisms, there is a general trend in the deviation of EF from 0.5 for the six compounds. First, the EFs in air are close to 0.5 in the abiotic compartments; second, the

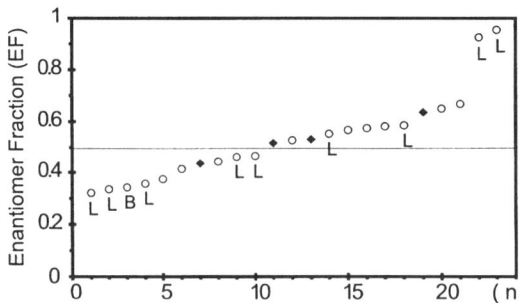

Fig. 7. Average enantiomer fraction of oxychlordane (OXY) in soil (◆) and biotic compartments (○) in ascending order by compartment number (n). L, liver tissue; B, brain tissue.

EFs in water compartments deviate slightly from 0.5. The highest deviations from EF = 0.5 in abiotic compartments occurred in soils. For the compounds considered, the order of the deviation from EF = 0.5 is air < water < soil. Enantioselective biodegradation of pesticides has a stronger effect on the EF in soils than in air or water compartments because pesticides from other sources with EFs close to 0.5 exchange easily between air and water. Also, compounds in a compartment with EFs close to 0.5 in a huge reservoir like an ocean are not affected by minor changes in EFs from other sources.

In general, biotic compartments (whole organisms, tissues, and organs) showed a higher deviation from EF = 0.5 than abiotic compartments; this is attributed to stereospecific metabolization and enzymatic transport processes. For the different compounds studied, lower trophic biota (e.g., mussels, cod, flounder) showed a smaller deviation from EF = 0.5 than higher trophic organisms (e.g., seals, birds, and terrestrial animals), as was also the case with α-HCH and octachlordane in the polar bear food chain (Wiberg et al. 1998a). A possible reason for these differences is that higher trophic level organisms have a higher metabolic capacity than cold water organisms (Karlsson et al. 2000).

High deviations from EF = 0.5 were found for the different pesticides in specific organs in biota, for example, in liver, kidney, brain tissue, and spinal marrow (Figs. 2, 4, 5, 6, 7). For the biotic compartments, a general trend in the deviation of EF = 0.5 was found in the following sequence: lower trophic biota < higher trophic biota < liver/kidney < brain. This sequence is the result of the combined effect of stereoselective degradation/metabolization, complexation, uptake, and excretion within an organism or its organ. Unfortunately, the mechanistic molecular processes are not understood, but progress has been made in understanding the mechanisms of chiral drugs applied to humans (Rochat et al. 1999).

B. Constant Enantiomer Fraction

After application, chiral xenobiotic compounds in the environment are redistributed over different compartments. In general, uptake and release of hydrophobic compounds are controlled by kinetic processes (de Wolf 1992) and hydrophobic compounds biomagnify from lower trophic to higher trophic organisms (de Voogt 1996).

When hydrophobic chiral compounds with EFs close to 0.5 are outside an organism, they can exchange with tissue and organs inside the organism. If no enantioselective processes occurred, the EFs of chiral compounds would have the same values inside as outside the organisms; however, enantioselective processes in different organs or tissues push the EF away from 0.5. The EF represents the result of these enantioselective exchange processes, enantioselective metabolization processes, and nonenantioselective exchange processes. Therefore, every organism and every organ in an organism will obtain its constant EF for a chiral compound that is typical for the compartment selected. In fact, the observed concentration of enantiomers is a combined effect of nonenantioselective exchange and enantioselective processes in which every biotic compartment attains a constant EF. In these processes, it is not important how the EFs of a compounds are achieved; they can be the result of several processes, such as enantiomer-selective transport, enantioselective metabolization, kinetic exchange, or food web accumulation.

C. Model Scheme

Deviations from EF = 0.5 in abiotic and biotic compartments were found for the six selected pesticides but also occurred for other chiral compounds released to the natural environment. In Fig. 8, we present a general scheme for the deviation of EF from the racemic mixture that is applicable to various chiral compounds. Nearly every chiral compound is induced in the environment as a racemate (EF = 0.5) as a result of a nonenantioselective chemical synthesis. Enantioselective processes are responsible for the deviations from EF = 0.5. The EFs of chiral compounds in the air, water, and soil compartments are close to the EF of the racemic mixture (Fig. 8). In organisms such as mollusks, fish, birds, and marine mammals the deviations from EF = 0.5 increase as a result of the active uptake from lower trophic organisms and further enantiomer enrichment. The strong enrichment of one enantiomer in a higher trophic level overrules that of the existing EF in a lower trophic organism, even if the enrichment favors the opposite enantiomer in the lower trophic level. This mechanism has been demonstrated in a polar bear food chain (Wiberg et al. 1998a) for α-HCH where cod showed enantioselective behavior opposite to that of the predatory ringed

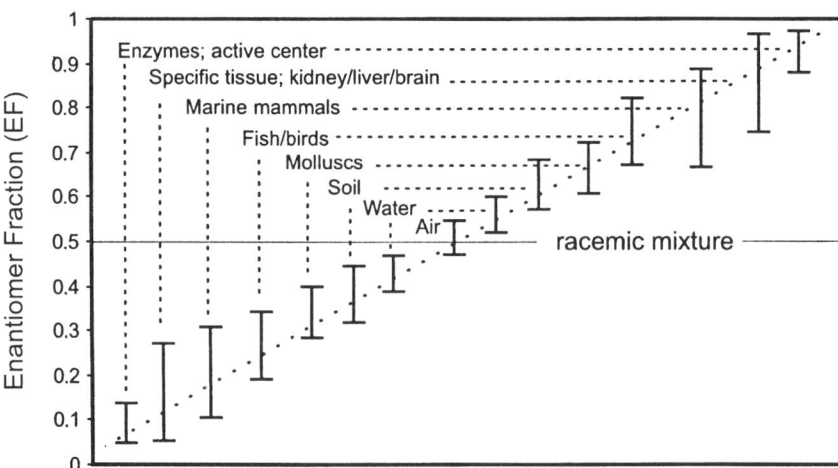

Fig. 8. Hypothetical model of enantiomer fractions (EFs) for a chiral compound in different compartments. Deviations from the racemic mixture are above or below the line EF = 0.5.

seal. However, by contrast, toxaphene compounds showed enantiomer enrichment in the lower trophic levels and racemic composition in the higher food web levels (Vetter and Maruya 2000).

The enrichment within a compartment shows a preference for one enantiomer; the EF is either above or below 0.5. (Fig. 8). The highest deviation of EF from the racemic mixture can be observed in liver, kidney, and brain tissue because in these compartments many stereoselective processes are in operation, i.e., enzymatic degradation and stereoselective transportation across membranes. On a molecular scale, we expect very high deviations of EF = 0.5 in the active centers of enzymes (Fig. 8) because this is where the chemical recognition of the chiral compound takes place. However, no data are available concerning enantiomer enrichment from enzyme active centers. Nonenantioselective biomagnification of a hydrophobic compound in a higher trophic organism will not alter the enantiomeric ratio; only enantioselective metabolization or enantioselective uptake or release can alter the EF. Although the processes leading to enantiomeric enrichment are not completely understood (Dugas and Penney 1981), we expect that the deviation from EF = 0.5 for various chiral compounds entering the environment will generally follow the scheme in Fig. 8.

D. Stereochemical Recognition of Chlordane Compounds

The structures of *cis*-chlordane (CC) and *trans*-chlordane (TC) in Fig. 1 show that these compounds differ only in the position of one Cl atom (R_1 versus R_2). These compounds enter the environment as racemic mixtures. If the stereochemical behavior is the same for each of these compounds, i.e., each compound is attacked at a topological congruent position, then we expect the deviation from EF = 0.5 to be in the same direction. In comparable compartments, we expect the EF to be either higher or lower than 0.5 for CC and TC. The direction of the deviation from the racemic mixture is shown in Table 3 for comparable compartments. CC and TC showed opposite behavior for most comparable compartments, which indicates opposite stereochemical behavior. The R_1 or R_2 position of the Cl atom (Fig. 1) proved to be the important structural position.

The structures of heptachlor *exo*-epoxide (HEPX) and oxychlordane (OXY) in Fig. 1 differ only in the R_3-substituent (Cl or H). These compounds enter the environment through the metabolism of chlordane compounds (Müller and Buser 1994). The direction of the deviation from EF = 0.5 (see Table 3) does not allow a similar conclusion to be drawn with regard to the stereochemical behavior of CC and TC. In some compartments similar stereochemical behavior was observed for HEPX and OXY whereas in other compartments opposite stereochemical behavior was observed (Table 3). On the molecular scale, the precise nature of the metabolization is complicated. The tertiary structure of a protein may be responsible for the stereochemical recognition of compounds and not the reaction center itself (Kaye 1991). However, certain structural elements can influence the observed enantiomer enrichment (Karlsson et al. 2000).

Table 3. Enantiospecific processes for the average enantiomer fraction (EF) of *cis*-chlordane (CC), *trans*-chlordane (TC), heptachlor *exo*-epoxide (HEPX), and oxychlordane (OXY) in soil, air, water, and biotic compartments.

Compartment	CC	TC	HEPX	OXY
Agricultural soils[a]	+	—	+	+
Air[b]	+	—	+	n.a.
Arctic water	+	—	+	n.a.
Grey seal, liver	—	—	n.a.	—
Grey seal, blubber	—	—	—	+
Harbor seal, blubber	—	+	—	—
Ringed seal, liver	+	—	n.a.	—
Ringed seal, blubber	+	—	—	+
Baltic herring, oil	+	—	n.a.	n.a.
Herring, total	+	—	—	+
Baltic salmon	—	+	n.a.	n.a.
Cod, liver oil	—	+	+	—
Seagull, egg	n.a.	n.a.	+	+
Hare, liver	n.a.	n.a.	+	+
Roe-deer, liver	n.a.	n.a.	+	+

Enantiomer-specific process: +, EF ≥ 0.5; —, EF < 0.5.
n.a., not available.
[a]CC and TC, 7 comparable compartments; HEPX and OXY, 3 comparable compartments.
[b]CC and TC, 3 comparable compartments; HEPX, 3 compartments.
Note: only grey seal compartments had the same sign for CC and TC.

E. Shielding from the Racemate

A deviation from EF = 0.5 can occur only by chemical binding of compounds where the enantiomers form diastereomers or by the formation of diastereomeric associations in which the interactions are weaker. Living systems, which range from simple to complex organisms, are homochiral and provide diastereomeric associations in enzymes that may result in stereoselective metabolization. Also, chiral compounds can be stereoselectively transported through biological membranes. For a deviation from EF = 0.5 to occur in any compartment, enantiomers have to be separated in a physical or a biochemical way to avoid a selective exchange with compartments that are not enriched in enantiomers. This "shielding from the racemate" can be obtained via membranes in living organisms, via semiisolated water bodies, by movements of percolating groundwater, or by sedimentation. Shielding of organisms from the surrounding water (EF ≈ 0.5) may be the result of the metabolism of chiral compounds and the selective passage of chiral compounds through biological membranes. This passage through biological membranes, the mechanisms of which are not completely understood (Baudry et al. 1997), can lead to different EFs within compartments.

For example, the EF of a organism can differ from the EF of water or the EF of liver can differ from the EF of brain (Figs. 2, 4, 5, 6, 7). In Arctic water, deep water is isolated from exchange processes with the atmosphere by overlying water bodies. In the Arctic atmosphere, the EF of α-HCH is approximately 0.5 whereas in subsurface water EFs of α-HCH reached 0.17–0.23 at depths of 250–1000 m (Harner et al. 1999).

Enantioselective processes in microorganisms may also alter the EF considerably (Harner et al. 1999). For instance, percolation water from waste disposal sites polluted with mecoprop has revealed EFs as high as 0.88 as a result of the groundwater being cut off from the racemic bulk by its movement in the ground (Zipper et al. 1998a). Shielding effects can be expected in sediment cores where each sediment layer is cut off from layers above or below. Only small changes in EF have been reported in sediment cores in which enantiomers of toxaphene compounds were measured (Vetter et al. 1998, 1999). The EFs of heptachlorobornane changed from 0.41 (in 1992) to approximately 0.44 (in 1935). The EFs of hepatchlorbornane were about 0.5 in all sediment layers. Changes in EFs in sediment cores were attributed to different types of microorganisms in different layers (Vetter et al. 1998, 1999).

F. Chiral Enrichment Processes

The extreme EFs depicted in Figs. 2 to 7 are found almost exclusively in the brain tissue of various organisms. Hühnerfuss et al. (1992a) attributed this high enrichment of one enantiomer relative to the other to the blood–brain barrier (BBB). BBB transport processes are currently under study by pharmacologists who seek to understand the action of chiral drugs (Rochat et al. 1999). The microvessels of the BBB consist of a single continuous layer of cerebral endothelial cells that are effectively sealed together by tight intercellular junctions (Ghersi-Egea et al. 1995). These junctions eliminate the paracellular pathway of solute movement through the BBB; there is practically no transcellular bulk flow of solute through the BBB (Baudry et al. 1997). Rochat et al. (1999) recognized three transport routes through the BBB: (1) passive passage, bidirectional, saturable transport that is characteristic of a mechanism requiring no energy; (2) carrier-mediated, by the presence of the efflux pump P-glycoprotein (Pgp) at the apical membrane of brain endothelial cells which effluxes many drugs and peptides back into the blood; and (3) metabolism, as brain endothelial cells contain a substantial volume of mitochondria, indicating that the BBB could contribute to significant monoamine oxidase (MAO) biotransformation of xenobiotics. Carrier-mediated transport and metabolism may result in the stereoselective passage of chiral compounds through the BBB, which in turn could lead to the observed changes in enantiomer fractions. If a chiral compound is to pass stereoselectively through a membrane, it has to create diastereomeric associations or complexes with its host, the membrane. Therefore, passive passage will not discriminate between enantiomers because there is practically no interaction with the membrane. Enantiomeric enrichments can be observed in liver samples (see Figs. 2, 4, 5, 6, 7). The liver is a detoxification organ for xenobiotics. The metabolization of atropisomers of polychlorobiphenyls (PCBs) in human liver samples produced exclusively one atropisomer of methylsulfonyl PCBs (Ellerichmann et al. 1998; Bergman et al. 1998).

We suggest that two processes lead to chiral enrichment, resulting in higher concentrations of one enantiomer: the first process is stereoselective degradation and the second is stereoselective separation. We refer here to the first process as the chiral machine and to the second as the chiral guard. The chiral machine denotes the enzymatic recognition and metabolization of chiral compounds. The chiral guard is nondestructive, and its principal process is chiral recognition by the endothelial cells and stereochemical effluxing of the chiral compound by Pgp. An EF higher or lower than 0.5 is the result of one or more stereospecific processes. We assume that in the liver the chiral machine is the main stereospecific process. The chiral enrichment in the brain is the result of the chiral guard in combination with the chiral machine.

G. Environmental Regulations

The EFs of several chiral compounds show that biotic processes treat enantiomers differently, resulting in unequal enantiomer concentrations even in abiotic compartments such as water, air, and soil. Enantiomers can have different biological and physiological properties and can affect plants and organisms in different ways. Therefore, investigations that treat racemates as though they were single entities can produce inaccurate and misleading results (Amstrong et al. 1993). In the U.S., almost all environmental regulations, which are based on toxicological studies, treat racemates as single molecules with the same properties (Schneiderheinze et al. 1999; Kohler et al. 1997). In the European Union, also, the possible different effects of enantiomers are not taken into account. Further research into enantioselective biodegradation and toxicology is needed to investigate the distribution effect and fate of chiral agrochemicals (Schneiderheinze et al. 1999). Therefore, toxicological studies for chiral agrochemicals do not have a sound scientific basis.

Summary

Enantiomer fractions (EFs) of chiral compounds have been used to explain the mechanisms of enantiomer enrichment in air, soil, water, and biota. The EFs were calculated from enantiomeric ratios (ERs) of chiral compounds measured by researchers during the past 10 years. Six compounds were selected from different abiotic and biotic compartments: α-hexachlorocyclohexane (α-HCH), mecoprop, *cis*-chlordane (CC), *trans*-chlordane (TC), heptachlor *exo*-epoxide (HEPX), and oxychlordane (OXY). The EF was used as a general descriptor for enantiomer enrichment. In environmental compartments the EFs of chiral pesticides deviated from those of the racemic composition (EF = 0.5). The deviations from EF = 0.5 in the different compartments show similar patterns for several compounds, i.e., air < water < soil < biota. In biota the order was lower trophic level < higher trophic level and liver or kidney tissue < brain tissue. Explanations for stereoselective behavior were found in pharmacology and brain research. The enantiomeric enrichments in environmental compartments were visualized in a general scheme applicable to other persistent chiral compounds. The mechanisms of enantiomer enrichment were conceptualized by a hypothetical model of a chiral machine (enzymatic degradation) and a chiral guard (stereospecific efflux). Environmental regulation authorities should treat chiral pesticides as a composition of enantiomers because biotic processes handle enantiomers as separate chemical entities.

Appendix. Chemical nomenclature.

Common name (abbreviation)	IUPAC	CAS number
HCH	Technical hexachlorocyclohexane	608-73-1
rac. (±)-α-HCH	(1a,2a,3b,4a,5b,6b)-Hexachlorocyclohexane	319-84-6
(+)-α-HCH	(1a,2a,3b,4a,5b,6b)-Hexachlorocyclohexane	119911-69-2
(−)-α-HCH	(1a,2a,3b,4a,5b,6b)-Hexachlorocyclohexane	119911-70-5
rac. (±)-(RS)-Mecoprop	(±)-(RS)-2-(4-Chloro-2-methylphenoxy)propanoic acid	7085-19-0
(+)-(R)-Mecoprop	(+)-(R)-2-(4-chloro-2-methylphenoxy)propanoic acid	16484-77-8
rac. (±)-cis-Chlordane (CC)	1-exo-2-exo,4,5,6,7,8,8-Octachloro-3a,4,7,7a-tetrahydro-4,7-methanoindane	5103-71-9
rac. (±)-trans-Chlordane (TC)	1-exo-2-endo,4,5,6,7,8,8-octachloro-3a,4,7,7a-tetrahydro-4,7-methanoindane	5103-74-2
rac. (±)-Heptachlor exo-epoxide or cis-heptachlorepoxide (HEPX)	1-exo-4,5,6,7,8,8-Heptachloro-2,3-exo-epoxy-3a,4,7,7a-tetrahydro-4,7-methanoindane	1024-57-3
rac. (±)-Oxychlordane (OXY)	1-exo,2-exo,4,5,6,7,8,8-Octachloro-2,4-exo-epoxy-3a,4,7,7a-tetrahydro-4,7-methanoindane	27304-13-8

rac., racemic.

Acknowledgments

This study was supported by the Dutch Ministry of Transport and Public Works, National Institute for Coastal and Marine Management/RIKZ. T.F. Bidleman, R.L. Falconer, W. Vetter, K. Wiberg, C. Zipper, H. Hühnerfuss, and K. Olie provided some of the literature. B.G. de Boer and F. Russel are thanked for supplying the BBB literature. The comments and corrections of P. de Voogt and J.C. Klamer improved the manuscript considerably. S.M. McNab gave attention to style and language. G.J.J. Groeneveld made the figures.

References

Aigner EJ, Leone AD, Falconer RL (1998) Concentrations and enantiomeric ratios of organochlorine pesticides in soils from the U.S. corn belt. Environ Sci Technol 32: 1162–1168.

Amstrong DW, Reid GL III, Hilton ML, Chang CD (1993) Relevance of enantiomeric separations in environmental science. Environ Pollut 79:51–58.

Aoyama T, Kotaki H, Sawada Y, Iga T (1994) Stereospecific distribution of methylphenidate enantiomers in rat brain: specific binding to dopamine reuptake sites. Pharm Res (NY) 11:407–411.

Aspeslet LJ, Baker GB, Coutts RT, Torok-Both GA (1994) The effects of desipramine and iprindole on levels of enantiomers of fluoxetine in rat brain and urine. Chirality 6:86–90.

ATSDR (Agency for Toxic Substances and Diseases Registry) (1994) Toxicology profile for chlordane (update). U.S. Department of Human and Health Services, Public Health Service, Atlanta, GA.

Baldessarini RJ, Kula NS, Zong R, Neumeyer JL (1994) Receptor affinites of aporphine enantiomers in rat brain tissue. Eur J Pharmacol 254:199–203.

Baudry S, Pham YT, Baune B, Vidrequin S, Crevoisier CH, Gimenez F, Farinotti R (1997) Stereoselective passage of mefloquine through the blood-brain barrier in the rat. J Pharm Pharmacol 49:1086–1090.

Bergman Å, Ellerichmann T, Franke S, Hühnerfuss H, Jakobsson E, König WA, Larsson C (1998) Gas chromatographic enantiomer separations of chiral PCB methyl sulfones and identification of selectively retained enantiomers in human liver. Organohalogen Compd 35:339–342.

Bertilsson L, Otani K, Dahl ML, Nordin C, Åberg-Wistedt A (1991) Stereoselective efflux of (E)-10-hydroxynortriptyline enantiomers from the cerebrospinal fluid of depressed patients. Pharmacol Toxicol 68:100–103.

Bidleman TF (1998) Transport and fate. Highlights of degradation and fate, air-surface exchange and physicochemical properties. Dioxin 1998, 17–21 August, Stockholm, Sweden.

Bidleman TF, Falconer RL (1999a) Using enantiomers to trace pesticide emissions. Environ Sci Technol 33:206A–209A.

Bidleman TF, Falconer RL (1999b) Enantiomer ratios for apportioning two sources of chiral compounds. Environ Sci Technol 33:2299–2301.

Bidleman TF, Jantunen LM, Harner T, Wiberg K, Wideman JL, Brice K, Su K, Falconer RL, Aigner EJ, Leone AD, Ridal JJ, Kerman B, Finizio A, Alegria H, Parkhurst WJ, Szeto SY (1998a) Chiral pesticides as tracers of air-surface exchange. Environ Pollut 102:43–49.

Bidleman TF, Jantunen LMM, Wiberg K, Harner T, Brice KA, Su K, Falconer RL, Leone AD, Aigner EJ, Parkhurst WJ (1998b) Soil as a source of atmospheric heptachlor epoxide. Environ Sci Technol 32:1546–1548.

Blanch GP, Glausch A, Schurig V, Serrano R, Gonzalez MJ (1996) Quantification and determination of enantiomeric ratios of chiral PCB 95, PCB 132, and PCB 149 in shark liver samples (*C. coelolepis*) from the Atlantic Ocean. J High Resolut Chromatogr 19:392–396.

Blaser H-U, Spindler F (1997) Enantioselective catalysis for agrochemicals: the case history of the Dual Magnum herbicide. Chimia 51:297–299.

Brossi A (1994) Chiral drugs: synopsis. Med Res Rev 14:665–691.

Bucheli TD, Müller SR, Heberle S, Schwartzenbach RP (1998a) Occurrence and behavior of pesticides in rainwater, roof runoff, and artificial stormwater infiltration. Environ Sci Technol 32:3457–3464.

Bucheli TD, Müller SR, Voegelin A, Schwartzenbach RP (1998b) Bituminous roof sealing membranes as major source of the herbicide (R,S)-Mecoprop in roof runoff waters: potential contamination of groundwater and surface waters. Environ Sci Technol 32:3465–3471.

Buser H-R, Müller MD (1993) Enantioselective determination of chlordane components, metabolites, and photoconversion products in environmental samples using chiral high-resolution gas chromatography and mass spectrometry. Environ Sci Technol 27: 1211–1220.

Buser H-R, Müller MD (1995a) Isomer-selective and enantiomerselective determination of DDT and related compounds using chiral high-resolution gas chromatography/ mass spectrometry and chiral high-performance liquid chromatography. Anal Chem 67:2691–2698.

Buser H-R, Müller MD (1995b) Isomer and enantioselective degradation of hexachlorocyclohexane isomers in sewage sludge under anaerobic conditions. Environ Sci Technol 29:664–672.

Buser H-R, Müller MD (1997) Conversion reactions of various phenoxyalkanoic acid herbicides in soil. 2. Elucidation of the enantiomerization process of chiral phenoxy acids from incubation in D_2O/soil system. Environ Sci Technol 31:1960–1967.

Buser H-R, Müller MD (1998) Occurrence and transformation reactions of chiral and achiral phenoxyalkanoic acid herbicides in lakes and rivers in Switzerland. Environ Sci Technol 32:626–633.

Buser H-R, Müller MD, Rappe C (1992) Enantioselective determination of chlordane components using chiral high-resolution gas chromatography-mass spectroscopy with application to environmental samples. Environ Sci Technol 26:1533–1540.

Buser H-R, Müller MD, Theobald N (1998) Occurence of the pharmaceutical drug Clofibric acid and the herbicide Mecoprop in various Swiss lakes and in the North Sea. Environ Sci Technol 32:188–192.

Caldwell J (1996) Importance of stereospecific bioanalytical monitoring in drug development. J Chromatogr A 719:3–13.

Christen K (1999) U.N. negotiations on POPs snag on malaria. Environ Sci Technol 33: 444A–445A.

CTB (1999) Annual report 1998. Dutch Board for Authorisation of Pesticides, Wageningen.

Deo PG, Karanth NG, Karanth NGK (1994) Biodegradation of hexachlorcyclohexane isomers in soil and food environment. Crit Rev Microbiol 20:57–78.

de Voogt P (1996) Ecotoxicology of chlorinated aromatic hydrocarbons. In: Hester RE, Harrison RM (eds) Chlorinated Organic Micropollutants. Issues in Environmental Science and Technology, Number 6. The Royal Society of Chemistry, Cambridge, UK, pp 89–112.

de Wolf W (1992) Influence of biotransformation on the bioconcentration of chemicals in fish. Ph.D. Thesis, Utrecht University, Utrecht.

Dugas H, Penney C (1981) Bioorganic Chemistry. A Chemical Approach to Enzyme Action. Springer-Verlag, New York, pp 267–282.

Eckhardt W, Francotte E, Herzog J, Margot P, Rihs G, Kunz W (1992) Synthesis and fungicidal activities of CGA 80000 (α-[N-(3-chloro-2,6-xylyl)-2-methoxyacetamido]-gamma-butyrolactone) and of its four isomers. Pestic Sci 36:223–232.

Ellerichmann T, Bergman Å, Franke S, Hühnerfuss H, Jakobsson E, König WA, Larson C (1998) Gas chromatographic enantiomer separations of chiral PCB methyl sulfones and identification of selectively retained enantiomers in human liver. Fresenius Environ Bull 7:244–257.

El Rassi Z (1997) Capillary electrophoresis of pesticides. Electrophoresis 18:2465–2481.

Falconer RL, Bidleman TF, Gregor DJ, Semkin R, Teixeira C (1995a) Enantioselective breakdown of α-hexachlorocyclohexane in a small arctic lake and its watershed. Environ Sci Technol 29:1297–1302.

Falconer RL, Bidleman, TF, Gregor DJ (1995b) Air-water gas exchange and evidence for metabolism of hexachlorocyclohexanes in Resolute Bay, N.W.T. Sci Total Environ 160/161:65–74.

Falconer RL, Bidleman TF, Szeto SY (1997) Chiral pesticides in soils of the Fraser Valley, British Columbia. J Agric Food Chem 45:1946–1951.

Falconer R, Leone A, Bodnar C, Wiberg K, Bidleman T, Jantunen L, Harner T, Parkhurst W, Alegria H, Brice K, Su K (1998) Using enantiomeric ratios to determine sources of chlordane to ambient air. Organohalogen Compd 35:331–334.

Faller J, Hühnerfuss H, König WA, Krebber R, Ludwig P (1991a) Do marine bacteria degrade α-hexachlorocyclohexane stereoselectively? Environ Sci Technol 25:676–678.

Faller J, Hühnerfuss H, König WA, Ludwig P (1991b) Gas chromatographic separation of the enantiomers of marine organic pollutants. Distribution of α-HCH enantiomers in the North Sea. Mar Pollut Bull 22:82–86.

Felding G (1995) Leaching of phenoxyalkanoic acid herbicides from farmland. Sci Total Environ 168:11–18.

Fielding M, Barcelo D, Helweg A, Galassi S, Torstensson L, Van Zoonen P, Wolter R, Angeletti G (1991) Pesticides in ground and drinking water. Review 27, Commission of the European Communities, Brussels.

Finizio A, Bidleman TF, Szeto SY (1998) Emission of chiral pesticides from an agricultural soil in the Fraser valley, British Columbia. Chemosphere 36:345–355.

Fuller RW, Snoddy HD (1993) Drug concentrations in mouse brain at pharmacologically active dose of fluoxetine enantiomers. Biochem Pharmacol 45:2355–2358.

Garrison AW, Schmitt P, Martens D, Kettrup A (1996) Enantiomeric selectivity in the environmental degradation of dichlorprop as determined by high-performance capillary electrophoresis. Environ Sci Technol 30:2449–2455.

Ghersi-Egea JF, Leininger-Muller B, Cecchelli R, Fenstermacher JD (1995) Blood-brain interfaces: relevance to cerebral drug metabolism. Toxicol Lett 82/83:645–653.

Gintautas PA, Daniel SR, Macalady DL (1992) Phenoxyalkanoic acid herbicides in municipal landfill leachates. Environ Sci Technol 26:517–521.

Glausch A, Hahn J, Schurig V (1995) Enantioselective determination of chiral 2,2′,3,3′,4,6′-hexachlorobiphenyl (PCB 132) in human milk samples by multidimensional gas chromatography/electron capture detection and by mass spectrometry. Chemosphere 30:2079–2085.

Haglund P, Wiberg K (1996) Determination of gas chromatographic elution sequences of the (+)- and (−)-enantiomers of stable atropisomeric PCBs on Chirasil-Dex. J High Resol Chromatogr 19:373–376.

Harner T, Kylin H, Bidleman TF, Strachan WMJ (1999) Removal of α- and γ-hexachlorocyclohexane and enantiomers of α-hexacyclohexane in the Eastern Arctic Ocean. Environ Sci Technol 33:1157–1164.

Harner T, Wiberg K, Norstrom R (2000) Enantiomer fractions are preferred to enanti-

omer ratios for describing signatures in environmental analysis. Environ Sci Technol 34:218–220.
Hashimoto A, Nishikawa T, Oka T, Takahashi K, Hayashi T (1992) Determination of free amino acid enantiomers in rat brain and serum by high-performance liquid chromatography after derivatization with *N-tert*-butylozycarbonyl-L-cysteine and *o*-phthaldialdehyde. J Chromatogr 582:41–48.
Hühnerfuss H (2000) Chromatographic enantiomer separation of chiral xenobiotics and their metabolites—a versatile tool for process studies in marine and terrestrial ecosystems. Chemosphere 40:913–919.
Hühnerfuss H, Kallenborn R, König WA, Rimkus G (1992a) Preferential enrichment of the (+)-α-hexachlorocyclohexane enantiomer in cerebral matter of harbour seals. Organohalogen Compd 10:97–100.
Hühnerfuss H, Faller J, König WA, Ludwig P (1992b) Gas chromatographic separation of the enantiomers of marine pollutants. 4. Fate of hexachlorocyclohexane isomers in the Baltic and North Sea. Environ Sci Technol 26:2127–2133.
Hühnerfuss H, Faller J, Kallenborn R, König WA, Ludwig P, Pfaffenberger B, Oehme M, Rimkus G (1993) Enantioselective and nonenantioselective degradation of organic pollutants in the marine ecosystem. Chirality 5:393–399.
Hühnerfuss H, Pfaffenberger B, Gehrcke B, Karbe L, König WA, Landgraff O (1995) Stereochemical effects of PCBs in the marine environment: seasonal variation of coplanar and atropisomeric PCBs in blue mussels (*Mytilus edulis* L.) of the German Bight. Mar Pollut Bull 30:332–340.
Hummert K, Vetter W, Luckas B (1995) Levels of alpha-HCH, lindane, and enantiomeric ratios of alpha-HCH in marine mammals from the northern hemisphere. Chemosphere 31:3489–3500.
Iwata H, Tanabe S, Sakai N, Tatsukawa R (1993) Distribution of persistent organochlorines in the oceanic air and surface seawater and the role of oceans on their global transport and fate. Environ Sci Technol 27:1080–1098.
Iwata H, Tanabe S, Iida T, Baba N, Ludwig JP, Tatsukawa R (1998) Enantioselective accumulation of α-hexachlorocyclohexane in northern fur seals and double-crested cormorants: effects of biological and ecological factors in the higher trophic levels. Environ Sci Technol 32:2244–2249.
Jantunen LM, Bidleman TF (1996) Air-water gas exchange of hexachlorocyclohexanes (HCHs) and the enantiomers of α-HCH in arctic regions. J Geophys Res 101:28837–28846.
Jantunen LM, Bidleman TF (1997) Correction to: Air-water gas exchange of hexachlorocyclohexanes (HCHs) and the enantiomers of α-HCH in arctic regions. J Geophys Res 102:19279–19282.
Jantunen LMM, Bidleman TF (1998) Organochlorine pesticides and enantiomers of chiral pesticides in arctic ocean water. Arch Environ Contam Toxicol 35:218–228.
Jantunen LM, Kylin H, Bidleman TF (1998) Air-water gas exchange of hexachlorocyclohexanes and the enantiomeric ratios of α-HCH in the South Atlantic Ocean and Antarctica. Organohalogen Compd 35:347–350.
Kaye B (1991) Chiral drug metabolism; a perspective. Biochem Soc Trans 19:456–459.
Kallenborn R, Hühnerfuss H, König WA (1991) Enantioselective metabolism of (±)-α-1,2,3,4,5,6-hexachlorocyclohexane in organs of the eider duck. Angew Chem Int Ed Engl 30:320–321.
Kallenborn R, Oehme M, Vetter W, Parlar H (1994) Enantiomer selective separation of

toxaphene congeners isolated from seal blubber and obtained by synthesis. Chemosphere 28:89–98.

Kallenborn R, Planting S, Haugen J-E, Nybø S (1998) Congener-, isomer- and enantiomer-specific distribution or organochlorines in dippers (*Cinclus cinclus* L.) from southern Norway. Chemosphere 37:2489–2499.

Karlsson H, Oehme M, Skopp S, Burkow IC (2000) Enantiomer ratios of chlordane congeners are gender specific in cod (*Gadus morhua*) from the Barents Sea. Environ Sci Technol 34:2126–2130.

Klobes U, Vetter W, Luckas B, Skirnisson K, Plötz J (1998a) Levels and enantiomeric ratios of α-HCH, oxychlordane, and PCB 149 in blubber of harbour seals (*Phoca vitulina*) and grey seals (*Halichoerus grypus*) from Iceland and further species. Chemosphere 37:2501–2512.

Klobes U, Vetter W, Luckas B, Hottinger G (1998b) Enantioselective determination of 2-*endo*,3-*exo*,5-*endo*,6-*exo*-8,8,9,10-octachlorobornane (B8-1412) in environmental samples. Organohalogen Compd 35:359–362.

Kohler EH-P, Angst W, Giger W, Kanz C, Müller S, Suter MJ-F (1997) Environmental fate of chiral pollutants—the necessity of considering stereochemistry. Chimia 51: 947–951.

König WA, Icheln D, Runge T, Pfaffenberger B, Ludwig P, Hühnerfuss H (1991) Gas chromatographic enantiomer separation of agrochemicals using modified cyclodextrines. J High Resol Chromatogr 14:530–536.

König WA, Hardt IH, Gehrcke B, Hochmuth DH, Hühnerfuss H, Pfaffenberger B, Rimkus G (1994) Optically active reference compounds for environmental analysis obtained by preparative enantioselective gas chromatography. Angew Chem Int Ed Engl 33:2085–2087.

LaManna JC, Harrington JF, Vendel LM, Abi-Saleh K, Lust, WD, Harik SI (1993) Regional blood-brain lactate influx. Brain Res 614:164–170.

Malaiyandi M, Shah SM (1984) Evidence of photoisomerization of hexachlorocyclohexane isomers in the ecosphere. J Environ Sci Health A19:887–910.

March J (1985) Stereochemistry. In: March J (ed) Advanced Organic Chemistry. Wiley, New York, pp 86–150.

McErlane KM, Axelson J, Vaughan R, Kerr CR, Price JD, Igwemezie L, Pillai G (1990) Stereoselective pharmacokinetics of tocainide in human uraemic patients and in healthy subjects. Eur J Clin Pharmacol 39:373–376.

Miyazaki A, Hotta T, Marumo S, Sakai M (1978) Synthesis, absolute stereochemistry, and biological activity of optically active cyclodiene insecticides J Agric Food Chem 26:975–977.

Miyazaki A, Sakai M, Marumo S (1980) Synthesis and biological activity of optically active heptachlor, 2-chloroheptachlor, and 3-chloroheptachlor. J Agric Food Chem 28:1310–1311.

Moser H, Rihs G, Sauter H (1982) Der Einfluss von Atropisomerie und chiralem Zentrum auf die biologische Aktivität des Metolachlor. Z Naturforsch 87B:451–462.

Möller K, Hühnerfuss H, Rimkus G (1993) On the diversity of enzymatic degradation pathways of α-hexachlorocyclohexane as determined by chiral chromatography. J High Resol Chromatogr 16:672–673.

Mössner S, Spraker TR, Becker PR, Ballschmiter K (1992) Ratios of enantiomers of alpha-HCH and determination of alpha-, beta-, and gamma-HCH isomers in brain and other tissues of neonatal northern fur seals (*Callorhinus ursinus*). Chemosphere 24: 1171–1180.

Müller MD, Buser H-R (1994) Identification of the (+)- and (−)-enantiomers of chiral chlordane compounds using chiral high-performance liquid chromatography/chiroptical detection and chiral high-resolution gas chromatography/mass spectrometry. Anal Chem 66:2155–2162.

Müller MD, Buser H-R, Rappe C (1997) Enantioselective determination of various chlordane components and metabolites using high-resolution gas chromatography with a β-cyclodextrin derivative as chiral selector and electron-capture negative ion mass spectrometry detection. Chemosphere 34: 2407–2417.

Müller MD, Schlabach M, Oehme M (1992) Fast and precise determination of α-hexachlorocyclohexane enantiomers in environmental samples using chiral high-resolution gas chromatography. Environ Sci Technol 26:566–569.

Nordin C, Bertilsson L (1995) Active hydroxymetabolites of antidepressants. Emphasis on E-10-hydroxy-nortriptyline. Clin Pharmacokinet 28:26–40.

Nybäck H, Halldin C, Åhlin A, Curvall M, Eriksson L (1994) PET studies of the uptake of (S)- and (R)-[^{11}C]nicotine in the human brain: difficulties in visualizing specific receptor binding in vivo. Psychopharmacology 115:31–36.

Oehme M, Schlabach M, Hummert K, Luckas B, Nordøy ES (1995) Determination of levels of polychlorinated dibenzo-p-dioxins, dibenzofurans, biphenyls and pesticides in harp seals from the Greenland Sea. Sci Total Environ 162:75–91.

Penmetsa KV, Leidy RB, Shea D (1997) Enantiomeric and isomeric separation of herbicides using cyclodextrin-modified capillary zone electrophoresis. J Chromatogr A 790:225–234.

Pfaffenberger B, Hühnerfuss H, Kallenborn R, Köhler-Günther A, König WA, Krüner G (1992) Chromatographic separation of the enantiomers of marine pollutants. Part 6: Comparison of the enantioselective degradation of α-hexachlorocyclohexane in marine biota and water. Chemosphere 25:719–725.

Pfaffenberger B, Hardt I, Hühnerfuss H, König WA, Rimkus G, Glausch A, Schurig V, Hahn J (1994a) Enantioselective degradation of α-hexachlorocyclohexane and cyclodiene insecticides in roe-deer liver samples from different regions of Germany. Chemosphere 29:1543–1554.

Pfaffenberger B, Hühnerfuss H, Gehrcke B, Hardt I, König WA, Rimkus G (1994b) Gas chromatographic separation of the enantiomers of bromocylen in fish samples. Chemosphere 29:1385–1391.

Prien D, Rehn D, Blaschke G (1997) Enantioselective biotransformation of the chiral antihistaminic drug dimethindene in humans and rats. Arzneim-Forsch (Drug Res) 47:653–658.

Püttmann M, Mannschreck A, Oesch F, Robertson L (1989) Chiral effects in the induction of drug-metabolizing enzymes using synthetic atropisomers of polychlorinated biphenyls (PCBs). Biochem Pharmacol 38:1345–1352.

Reich S, Jimenez B, Marsili L, Hernandez LM, Schurig V, Gonzalez MJ (1999) Congener-specific determination and enantiomeric ratios of chiral polychlorinated biphenyls in striped dolphins (*Stenella coeruleoalba*) from the Mediterranean Sea. Environ Sci Technol 33:1787–1793.

Reist M, Testa B, Carrupt P-A, Jung M, Schurig V (1995) Racemization, enantiomerization, diastereomerization, and epimerization: their meaning and pharmacological significance. Chirality 7:396–400.

Renner R (2000) Researchers point toward chiral chemistry as pollution cure. Environ Sci Technol 34:9A–10A.

Ridal JJ, Bidleman TF, Kerman BR, Fox ME, Strachan WMJ (1997) Enantiomers of α-hexachlorocyclohexane as tracer of air-water gas exchange in Lake Ontario. Environ Sci Technol 31:1940–1945.

Rochat B, Amey M, Van Gelderen H, Testa B, Baumann P (1995) Determination of the enantiomers of Citalopram, its demethylated and propionic acid metabolites in human plasma by chiral HPLC. Chirality 7:389–395.

Rochat B, Amey M, Gillett M, Meyer UA, Baumann P (1997) Identification of three cytochrome P450 isozymes involved in N-demethylation of citolopram enantiomers in human liver microsomes. Pharmacogenetics 7:1–10.

Rochat B, Baumann P, Audus KL (1999) Transport mechanisms for the antidepressant citalopram in brain microvessel endothelium. Brain Res 831:229–236.

Sam E, Sarre S, Michotte Y, Verbeke N (1997) Distribution of apomorphine enantiomers in plasma, brain tissue and striatal extracellular fluid. Eur J Pharmacol 329:9–15.

Scanley BE, Baldwin RM, Laruelle M, Al-Tikriti S, Zea-Ponce Y, Zoghbi S, Giddings SS, Charney DS, Hoffer PB, Wang S, Gao Y, Neumeyer JL, Innis RB (1994) Active and inactive enantiomers of 2-β-carbomethoxy-3β-(4-iodophenyl)tropane: comparison using homogenate binding and single proton emission computered tomographic imaging. Mol Pharmacol 45:136–141.

Schmitt P, Garrison AW, Freitag D, Kettrup A (1997) Application of cyclodextrin-modified micellar electrokinetic chromatography to the separation of selected neutral pesticides and their enantiomers. J Chromatogr A 792:419–429.

Schneiderheinze JM, Armstrong DW, Berthod A (1999) Plant and soil enantioselective biodegradation of racemic phenoxyalkanoic herbicides. Chirality 11:330–337.

Sevcik J, Lemr K, Stransky Z, Vecera T, Hlavac J (1997) Possible uses of micellar electrokinetic capillary chromatography and high-performance liquid chromatography for the chiral discrimination of some pyrethroids. Chirality 9:162–166.

Sidhu J, Priskorn M, Poulsen M, Segonzac A, Grollier G, Larson F (1997) Steady-state pharmacokinetics of the enantiomers of Citalopram and its metabolites in humans. Chirality 9:686–692.

Simonyi M (1984) On chiral drug action. Med Res Rev 4:359–413.

Slooff W, Matthijsen AJCM (eds) (1988) Integrated criteria document hexachlorocyclohexanes. RIVM report 758473011. National Institute of Public Health and the Environment, Bilthoven.

Smith DF, Peterson HN (1982) Stereoselective effect of tranylcypromine enantiomers on brain serotonin. Life Sci 31:2449–2454.

Spangler LA, Mikalojczyk M, Burdge EL, Kielbasinski P, Smith HC, Lyzwa P, Fisher JD, Omelanczuk J (1999) Synthesis and biological activity of enantiomeric pairs of phosophosulfonate herbicides. J Agric Food Chem 47:318–321.

Tanabe S, Kumaran P, Iwata H, Tatsukawa R, Miyazaki N (1996) Enantiomeric ratios of α-hexachlorocyclohexane in blubber of small cetaceans. Mar Pollut Bull 32:27–31.

Ternay AL Jr (1979) The stereochemistry of ring systems. In: Ternay AL Jr (ed) Contemporary Organic Chemistry. Saunders, Philadelphia, pp 245–287.

Tett VA, Willetts AJ, Lappin-Scott HM (1994) Enantioselective degradation of the herbicide Mecoprop [2-(2-methyl-4-chlorophenoxy)propionic acid] by mixed and pure bacterial cultures. FEMS Microbiol Ecol 14:191–200.

Tett VA, Willetts AJ, Lappin-Scott HM (1997) Biodegradation of the chlorophenoxy herbicide (R)-(+)-Mecoprop by *Alcaligenes denitrificans*. Biodegradation 8:43–52.

Ulrich EM, Hites RA (1998) Enantiomeric ratios of chlordane-related compounds in air near the Great Lakes. Environ Sci Technol 32:1870–1874.

Verschueren K (1996) Handbook of Environmental Data on Organic Chemicals. Van Nostrand Reinhold, New York.

Vetter W, Maruya KA (2000) Congener and enantioselective analysis of toxaphene in sediment and food web of a contaminated estuarine wetland. Environ Sci Technol 34: 1627–1635.

Vetter W, Schurig V (1997) Enantioselective determination of chiral organochlorine compounds in biota by gas chromatography on modified cyclodextrines. J Chromatogr A 774:143–175.

Vetter W, Klobes U, Hummert K, Luckas B (1997a) Gas chromatographic separation of chiral organochlorines on modified cyclodextrin phases and results of marine biota samples. J High Resol Chromatogr 20:85–93.

Vetter W, Klobes U, Luckas B, Hottinger G (1997b) Enantiomeric resolution of persistent compounds of technical toxaphene (CTTs) on t-butyldimethylsilylated β-cyclodextrin phases. Chromatographia 45:255–262.

Vetter W, Bartha R, Stern G, Tomy G (1998) Enantioselective determination of two major compounds of technical toxaphene in Canadian lake sediment cores from the last 60 years. Organohalogen Compd 35:343–346.

Vetter W, Bartha R, Stern G, Tomy G (1999) Enantioselective determination of two persistent chlorobornane congeners in sediment from a toxaphene-treated Yukon lake. Environ Toxicol Chem 18:2775–2781.

Wania F, Mackay D (1999) Global chemical fate of α-hexachlorocyclohexane. 2. Use of a global distribution model for mass balancing, source apportionment, and trend prediction. Environ Toxicol Chem 18:1400–1407.

Wiberg K, Jantunen LM, Harner T, Wideman JL, Bidleman TF, Brice K, Su K, Falconer RL, Leone AD, Parkhurst W, Alegria H (1997) Chlordane enantiomers as source markers in ambient air. Organohalogen Compd 33:209–213.

Wiberg K, Letcher R, Sandau C, Norstrom R, Tysklind M, Bidleman TF (1998a) Enantioselective analysis of organochlorines in the Arctic marine food chain: chiral biomagnification factors and relationships of enantiomeric ratios, chemical residues and biological data. Organohalogen Compd 35:371–374.

Wiberg K, Oehme M, Haglund P, Karlsson H, Olsson M, Rappe C (1998b) Enantioselective analysis of organochlorine pesticides in herring and seal from the Swedish marine environment. Mar Pollut Bull 36:345–353.

Wiberg K, Letcher R, Sandau C, Duffe J, Norstrom R, Haglund P, Bidleman TF (1998c) Enantioselective gas chromatography/mass spectrometry of methylsulfonyl PCBs with application to Arctic marine mammals. Anal Chem 70:3845–3852.

Wiberg K, Letcher RJ, Sandau CD, Norstrom RJ, Tysklind M, Bidleman TF (2000) The enantioselective bioaccumulation of chiral chlordane and α-HCH contaminants in the polar bear food chain. Environ Sci Technol 34:2668–2674.

Willett KL, Ulrich EM, Hites RA (1998) Different toxicity and environmental fates of hexachlorocyclohexane isomers. Environ Sci Technol 32:2197–2207.

Williams A (1996) Opportunities for chiral agrochemicals. Pestic Sci 46:3–9.

Worthing CR, Hance RJ (1991) The Pesticide Manual: A World Compendium, 9th Ed. British Crop Protection Council, Farnham, Great Britain.

Yu H, Liu Y, Li HB, Martin AR, Hacksell U, Lewander T (1997) Pharmacodynamic and pharmacokinetic studies in rats of S-8-(2-furyl) and R-8-phenyl-2-(di-n-propylamino)tetralin, two novel 5-HT$_{1A}$ receptor agonists in vitro with different properties in vivo. J Pharm Pharmacol 49:169–177.

Zipper C (1998) Microbial degradation and environmental fate of chiral phenoxyalkanoic acid herbicides. Ph.D. thesis, Swiss Federal Institute of Technology, Zurich.

Zipper C, Suter MJ-F, Haderlein SB, Gruhl M, Kohler H-PE (1998a) Changes in the enantiomeric ratio of (R)- to (S)-Mecoprop indicate in situ biodegradation of this chiral herbicide in a polluted aquifer. Environ Sci Technol 32:2070–2076.

Zipper C, Bolliger C, Fleischmann T, Suter MJ-F, Angst W, Müller, Kohler H-PE (1998b) Fate of herbicides Mecoprop, Dichlorprop, and 2,4-D in aerobic and anaerobic sewage sludge as determined by laboratory batch studies and enantiomer-specific analysis. In: Zipper C. Microbial degradation and environmental fate of chiral phenoxyalkanoic acid herbicides. Ph.D. thesis, Swiss Federal Institute of Technology, Zurich, pp 73–92.

Manuscript received November 24, 2000; accepted February 24, 2001.

RMS Titanic and the Emergence of New Concepts on Consortial Nature of Microbial Events

D. Roy Cullimore, Charles Pellegrino, and Lori Johnston

Contents

I. Introduction	117
II. Consortial Structures Within Rusticles	121
III. Rusticle Growth Rates	123
IV. Rusticle Relevance to Maritime Industries	126
V. Rusticle Biomass on the *RMS Titanic*	128
VI. Focused Accumulation Sites for Iron Within Rusticles	129
VII. Water Wells and Rusticles	130
VIII. Laboratory Evaluation of Corrosive Processes	131
IX. Environmental Cost of Covert Rusticle Growth	133
X. Conclusion	138
Summary	139
Acknowledgment	140
References	140

I. Introduction

The *RMS Titanic* sank in the early-morning hours of April 15, 1912, and created a storm of controversy that still causes ripples around the world today. Her discovery as a shattered hull in 1985 by Dr. Robert D. Ballard revealed that the stem and the stern were torn apart, with the bow section looking as though it were docked on the ocean floor while the stern lay torn, twisted, and folded a kilometer away. One of the discoveries seen at that time was the growth of rusticles all over the hull. The name 'rusticle' was coined by Ballard because these growths appeared to resemble rust-covered icicles hanging downward all around and inside the ship (Ballard 1987).

Investigation of rusticles revealed that they were complex in structure, were formed as microbially induced concretions (a form of living concrete), and were not composed of a single species of either an animal or plant but were a complex network of microbial consortia. Taxonomists have traditionally classified all living entities as separate species and envisage that these would all normally live as distinct entities reproducing only within their own species. The microbial world, however, does not commonly follow such a simplistic scheme, and most growths of living masses involve a multiplicity of species functioning within a durable

---

Communicated by George W. Ware.

D.R. Cullimore (✉)·C. Pellegrino·L. Johnston
Droycon Bioconcepts Inc., 3303 Grant Rd., Regina, Saskatchewan, Canada S4S 5H4

consortium as the social group. By analogy, Marcus Aurelius (1969a) summarized the advantages of functioning in a coordinated manner in his meditations:

> The intelligence of the universe is social. Accordingly it has made inferior things for the sake of the superior, and it has fitted the superior to one another. Thou seest how it is subordinated, co-ordinated and assigned to everything its proper proportion, and has brought together into concord with one another the things which are the best.

Our understanding of life has long been based on the concept of individual species of animals or plants, each being composed of cells having a common hereditary origin from a single zygote. These organisms, commonly function independently as living entities, each possessing a definable individual space but sharing a common and usually interdependent reproductive process. With the establishment of microorganisms over the past two centuries as a separate and definable group distinctly different from plants and animals, the concept of species, already well established for these higher organisms, was automatically adopted as the primary mechanism for the systematic differentiation of microorganisms. In so doing, the concepts intrinsic to systematics in microbiology became established around the need to isolate, describe, culture, and classify organisms that had a completely common set of characteristics and could be defined as consisting of a single strain of microorganism. To achieve this, the source habitat would have routinely been subjected to intense laboratory-based practices in which all other species had been eliminated so that only a single selected strain survived in any given culture. Thus, the whole classification of microorganisms has been built around the single-species concept.

During the past decade, there has been a growing realization that the rigid application of this concept has been a factor limiting the effective classification of all microbial species. For example, Sly (1995) prepared a report, on behalf of the International Union of Microbiological Societies and International Union of Biological Societies International Committee on Microbial Diversity, that expressed some of the inherent weaknesses in the present application of systematics in microbiology. In stating that "Less than 5% of these species (of microorganisms) have so far been described," there would clearly be a concern that the bulk of microbial species remain enigmas. Like other enigmas, understanding this difficult and confusing problem will require significant thought and investigation. Sly (1995) further indicated that "very little is known about the occurrence of uniqueness of most groups of microbial species." These statements reinforce concerns that the species concept, as it has been applied in microbiology, may not be functionally effective for the understanding of most members of the microbial kingdom.

Recent research conducted by the authors suggests that, on many occasions, the single-species concept may be at fault when applied to microbiology. The primary reasoning for this is that within the natural world microorganisms frequently function not independently but cooperatively as a part of multispecies consortia. Consortia are associations of multiple microbial species that are able to function in a synergistic manner. The functioning of the community is dependent on the contributions of each of the component strains within the consortium. This concept is beginning to have an impact in the medical industry, where there is now a growing realization that many more diseases than previously anticipated are induced by microorganisms. In the determination of the causes of chronic disease, microbial vectors are rarely considered, and there is today a growing recognition that many chronic diseases have their origins in microbial infections of the human body (Ewald 2000). There has not yet been any recognition that the types of consortial infestations seen on the *RMS Titanic* could parallel similar types of infestations in humans. In other words, looking at the pathogen as being but the tip of the iceberg, then its arrival at a site of infection is actually the result of a consortial activity involving many different species of microbes. We readily see only the rusticles that are obvious but not the jelly-like rusticles which are diffused throughout the site being infested; this has been a particular challenge in porous media saturated with groundwater (Cullimore 2000a).

From the present experiential knowledge of the authors, it would appear that there are a number of forms within which these consortia can function. The primary differentiation would be species dependency upon the consortium. Those aggregations in which the member species are essential for, and dependent upon, the functioning of the consortium could be termed vital consortia. In contrast, consortia whose members are transiently interdependent upon one another but can exist individually could be termed transient consortia. In many cases, it would be expected that not all of the component strains in a vital consortium could be cultured independently. However, in the transient consortia, there would be a higher probability that the strains could sustain an independent state. These observations contradict the well-established concept that a living "organism" must have arisen from a single cell in which all the characteristics applicable to all the tissues and structures generated by that organism are retained. In a vital consortium, it would be expected that all the strains essential to the formation of that type of consortium would probably relocate and form as a single community. For transient consortia, it is probable that the various strains involved in these formations may have arrived independently and formed the structure as a result of environmental factors at the site.

All consortia collectively generate structures that become discernible and definable as describable entities. Examples are given here of various obvious and covert living structures that all function in a consortial manner. In each example, it was recognized that the incumbent microbial species might function collectively in an integrated manner within their common habitat. This interdependence means that most strains cannot be easily isolated and cultured as individual species under laboratory conditions. In these studies, it has become evident that there is a need to recognize the nature of often complex and large microbial consortial structures as distinct forms. Consortia should be granted recognition as a separate growth form worthy of recognition. This recognition would be as a separate and definable group within the systems of biological classification. However, with the consortial growths, such species-based concepts cannot function adequately to formulate the origins of the consortium. In addition, there is the need to separately culture and describe the microbial strains recovered from a consortium. Once this has been achieved, then the reassemblage of the strains in the appropriate environmental setting should trigger the reestablishment of the consortia in both form and structure. This objective may be more achievable for a transient form of consortium than for a vital form.

Each example of a consortium discussed here has been subjected to microbiological investigation at both the structural and species composition levels. Individual species and groups of microorganisms have been isolated and, in some cases, the consortial structures have been reestablished. In each case, the generation of these structures involves the deliberate admission of a range of species into a favorable environment where they will jointly cause the formation of the typical form of growth. Examples of consortia include (a) rusticles recovered from the stem of *RMS Titanic*, (b) clog and tubercle formations infesting groundwater around water wells, and (c) effects of rusticle formation on the maritime industry.

In the evaluation of the environmental impacts that can be associated with the sinking of the *RMS Titanic*, possibly the key factor would be the nature and form of the microbial consortial activities occurring at the site. The primary impact would be the dramatic arrival of the shattered hull, incorporating approximately 31,800 metric tons (t) of iron, on a relatively flat section of the ocean floor. Sequential impacts that could affect the generation of microbial activity include the following:

1. Electrolytic activities associated with the final discharges of any electrical storage systems as a short-term effect and interaction between materials having different electrical potentials as a longer-term effect.
2. Dispersion of foods, beverages, and other organic materials released by the physical disruption of containers. These organics would principally originate from the refrigerated storage rooms in the lower decks and be approximately equivalent to 14 t carbohydrate, 4.5 t protein, and 4 t fat from that source alone.
3. Biodeterioration of the readily degradable organic materials used in the construction of the ship, including cellulosic material such as curtains, sheets, napkins, and paper.
4. Biodeterioration of the more recalcitrant organic materials such as the softwood structures.
5. Corrosion of the metallic structures, particularly those dominated by iron.
6. Structural collapse of the hull.

Each of these events would have a distinct impact on the level of activity of the biota colonizing at the site. For the purposes of this review, biota is considered to include forms of microbial life as well as the more traditionally recognized plants and animals. In the generation of a food web at the site, the microbial component may be expected to perform a key role through the sequential impacts of the effects just listed. In 1996, the most obvious elements of biological activity were the rattail fish, crabs, starfish, worms, and sea cucumbers. However, the largest biological component was the rusticle encrustation growing in various manners on the steel structures and gradually weakening the integrity of the ship. Scientific examination of these rusticles revealed that they were indeed generated by consortia of microbial species, feeding partly on the materials being released from the ship but also on the copious amounts of biocolloids and small creatures raining down from higher in the water column in the form of sea snow.

II. Consortial Structures Within Rusticles

From the brief descriptions given here, the case may be made that some living structures such as the rusticles are composed of a multiplicity of species rather than a single species. There would appear to be two levels of organizational structure involved in these consortia. One level would involve interdependence for the common survival of the consortium. In this instance, no one species would be able to physiologically or structurally maintain the structure without the activities of the other component species. The other level would involve the juxtapositioning of the species in relation to each other and also to their relative positioning in the physical structure within which the consortium lives. In many cases, these physical structures would include mechanisms to allow the entry of water, oxygen, and nutrients and passageways for the elimination of potential suppressive products of consortial activity (e.g., carbon dioxide, acidic metabolic products). Mechanisms would also be present to provide some protection to the consortial occupants from any physical disruption or predation. In the rusticles, this support is provided through the generation of geothite-dominated concretious structures, which are extremely porous. The presence of water channels, cavities, and extensive ductwork attest to the need of the consortium for water circulating through the bioconcretions. Microscopic examination at low power, at high power, and using scanning electron microscopy all revealed that complex structures are involved. In the more evolved biota, these complex structural differentiations could form a major factor in the development of specialized tissues and organs universal in these species. From the early observations, the types of extracellular structures observed in the rusticles appear to perform a function parallel to tissue differentiation in the more evolved biota. Possibly, it could be proposed that for microorganisms the formation of "tissues" was essentially extracellular; these could be organic polymer dominated, forming what are loosely described as "slimes." The slimes could come to contain inorganic crystallized and amorphous forms of insoluble substances. In the more highly evolved biota (i.e., plants and animals), the structures emerged largely in an intracellular manner. The classical perception has been to view tissue differentiation as forming the basis for the development of plants and animals with fungal structures being a major precursor route (e.g., differentiation of fruiting caps such as the mushroom *Agaricus campestris*). It is proposed here that the rusticles and parallel consortia represent an early stage in tissue differentiation primarily driven by extracellular process controlled by microbial consortia. Conversely, in the biota, the cells of the single species control tissue differentiation.

It is common to view the biota as being composed of a range of single-species organisms that operate independently, with few exceptions, and reproduce using some mechanism which causes like-structured organisms to be generated. Microorganisms do not follow this mandate; instead, a number of species function in a consortial manner to create a common habitat structure within which all the species cohabit. It has become apparent that structures incorporating microbial communities include that range of species essential to the basic functioning of the consortium. It clearly becomes impractical to isolate each of the species, particularly where a high level of interdependency has evolved.

It is therefore proposed that definable extracellular structures that have been formed out of, nurtured by, and matured as a result of the activities of an incumbent consortia of microbial species should be recognized and classified as a distinctly separate group within the living world. Where there are essentially a complex of different species forming populations that integrate into community structures, some form of "communication" must exist to allow mutualism. This exchange may be based partly upon the availability of space, electron acceptors, the redox potential, and, perhaps, upon some more direct form of cellular signal that may be chemical or physical in nature. This coordination of microbial behavior into complex patterns is known as "quorum sensing" (Straus 1997). The environmental impacts associated with the sinking of the *RMS Titanic* therefore relate to the stimulation of a myriad of consortial microbial activities that subsequently affect the rest of the local biota.

To this end, it is further proposed that a system of classification needs to be introduced in which the defining first word would be *consortium*; this would mark the living "organism" being described as consisting of a microbial consortium. Construction of the consortium is based on a minimum number of member species arriving at a suitable niche, which would then allow the consortium to generate a structure possessing the typical features expected for that form of growth (i.e., replication of a definable living growth as a result of a commonality of circumstances). These consortia would initiate the biodegradation of the structures and the relocation of any recalcitrant materials. The basic parameters that could be used to identify consortia at the community rather than at the individual or molecular level were defined by Cullimore (2000b). This approach involved culturing the sample within selective culture conditions including juxtaposition of diffusing selective nutrient and oxidization fronts to generate locational environmental patterns and reactions resulting from metabolic functions. The reactions and activities observed are formed into a reaction pattern sequence (RPS) that is typical for particular consortia. In generating RPS data, it now becomes possible to recognize the major bacterial types that may be contributing members of the consortium.

In the determination of environmental impacts, it has been a common practice to attempt to describe microbial activity primarily at the species and genome level because this is deemed to be intrinsically more accurate. However, such events do not recognize the importance of consortial activity as an essential component of the environmental process. For the rusticles on the *RMS Titanic*, the use of this level of precision would have proved expensive and inappropriate to the determination of the microbial components active within the rusticles. One significant outcome of the research on the *RMS Titanic* has been the recognition of the major importance of microbial consortia as the prime generators of the rusticles.

III. Rusticle Growth Rates

Since the discovery of the *RMS Titanic* in 1985, there have been a number of opportunities to examine this deep-sea wreck. It is, in fact, the world's first deep-ocean archeological site. The *Titanic* rests under approximately 4000 m of 1 °C water, with pressures in excess of 41,000 KPa. One of the first features that stood out in the images of the ship was the mass of rustlike growths (Pellegrino 2000), which occurred as a growing mass of iron-rich bioconcretions on the steel surfaces of this once-elegant ship. These bioconcretions were first noted growing on the outside and within the ship's structures. The scale of these growths led to the adoption of the term "rusticle" as a derivation of two words, "rust" and "icicle." The term "rust" was elected because the growths had a predominantly rusty color and the texture resembled flakes of rust growing on steel (Cullimore 1999). The most dramatic, but not necessarily the largest, growths hung over the sides of the hull, somewhat paralleling the structure of icicles.

In character and form, these growths resemble the speleothems that have been observed in natural limestone caves. These rare speleothems, originally considered to be secondary mineral growths, appear to be of subaqueous origin and show many similarities to the deep oceanic growths observed on the *RMS Titanic*. In their simplest form, both the rusticles and the speleothems could be described as elongate structures incorporating organic filaments coated by, or included in, a shell of iron oxide or calcite. In the Lechuguilla Cave in New Mexico, the speleothem growths are described as hanging down in a manner that closely resembles the rusticles at the *RMS Titanic* (Davis et al. 1990). These cave growths are irregular, consisting of iron oxide stalactites and calcite-encrusted columns; these are all of ancient origin (more than 100,000 yr ago) and are now dry and inactive. Microscopic examination reveals that the speleothems are primarily iron oxide deposits covering organic filaments. These encrustations were deposited via oxidative reactions that may have been initiated by bacteria. Other unusual cave features related to rusticles in form, but not based on iron oxides, include "pool fingers." These stalactiform subaqueous growths are calcite-encrusted organic strings, interconnected by curved bridge structures. In Wind Cave, South Dakota, hollow subaqueous calcite speleothems known as "helictite bushes" that grow in an upward branching pattern, have been identified. These growths again closely resemble rusticles and pool fingers; they include fossil bacterial traces but may be more closely related to the submarine "white smokers" than to either rusticles or the pool fingers (Davis 1989; Davis et al. 1990; LaRock and Cunningham 1995).

The bioconcretious rusticles vary in color, texture, size, and form. The variations of color, particularly the brilliant orange-brown color of the rusticles, are due to the highly oxidized ferric iron content. Closer examination of the rusticles by Pellegrino and Cullimore (1997), Wells and Mann (1997), and Mann (1997) revealed that the rusticles are complex structures involving water channels, reservoirs, complex iron platelike structures, threadlike spans, porous matrices, and ducts connecting to the outside. Within the rusticle structure appear to be a number of different microbial strains occupying specific sites. These strains were identified, using the Biological Activity Reaction Tests (BART^{TM}, Droycon Bioconcepts Inc., Canada), to include sulfate-reducing bacteria (SRB), iron-related bacteria (IRB), heterotrophic aerobic bacteria (HAB), denitrifying bacteria (DN), and archaeobacteria, together with a range of fungi. Six different forms of rusticles were noted; however, all bore the common characteristics of diverse and site-focused bacterial consortia (Fig. 1).

The supporting structures appeared to be dominated by a meshlike, heavily mineralized matrix in which goethite was dominant. The presence of goethite in rusticles was confirmed by Garzke et al. (1997). In addition, an iron oxide sulfate complex known as green rust [$Fe^{2+}_{3.6} Fe^{3+}_{0.9}(O^-, OH^-, SO_4^-)_9$] was found (Garzke et al. 1997). A large hanging rusticle recovered from the ship in 1996 was analyzed by electron diffraction x-ray, which revealed that iron was the dominant atom within the range of atoms tested. The relationship was (dominant atom first) Fe > Na > S > Cl > Ca > Mg > Si > P > Mn. There was a considerable variation in the elemental composition for the various samples analyzed, reflecting the heterogeneous nature of the structures within the rusticles. Where goethites dominate the structure, the iron (Fe) concentrations are very high whereas other components within the rusticle (e.g., the water channels and porous regions) have lower iron levels. Rusticles also vary widely in size and form.

Video imagery revealed rusticle sizes ranging from tiny tubercles or encrustations to massive, braided, or ropelike bioconcretions exceeding 3–4 m in length. The rusticle form also varied from flat, platelike growths to convoluted and intricate growth patterns. One major environmental impact of the sinking has therefore been the creation of a growing biomass dominated by rusticles that are gradually mobilizing the iron from the ship's steel into various oxidized forms which are moving away from the ship's structures to create associated biomass. Such an environmental impact has serious potential implications for the maritime industry.

The hull of the *RMS Titanic*, torn into three main sections, lies on the ocean floor in an oxidative environment where animal life can also be observed. In the deeper regions of the hull there is a strong probability that more reductive conditions may exist that would not only totally change the nature of the dominant microbial consortia but also influence the physical and chemical nature of the local environment. Gerhules and Alford (1990) observed such a phenome-

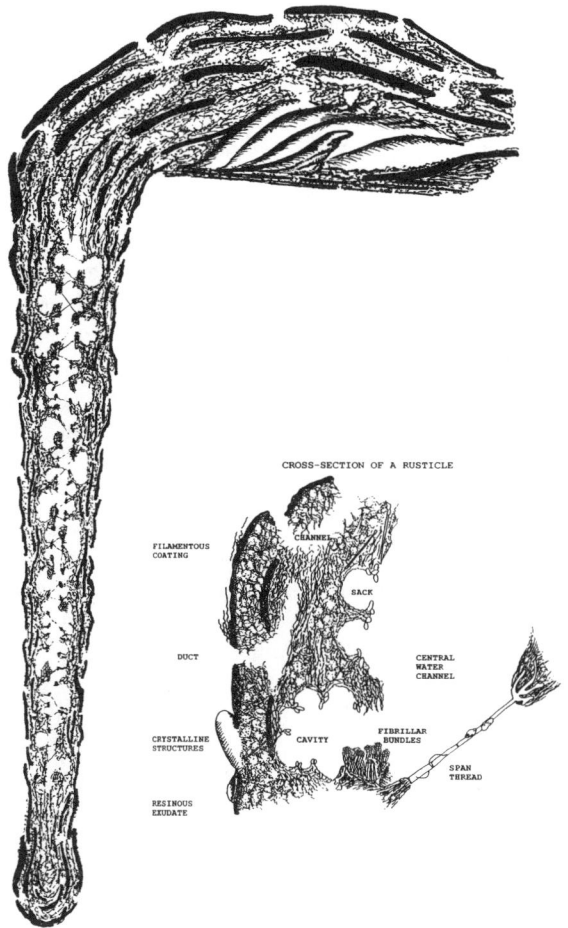

Fig. 1. Illustration of the typical components in a dissected hanging rusticle. The left side of the diagram is a vertical cross section through a hanging rusticle showing a central water channel with saclike extensions into the porous cortex of the rusticle. The outside of the rusticle is coated with iron-rich plates. To the lower right is an illustration of the structures found associated with the central water channel; these include the spans of threadlike materials forming a major structural support (upper left), the cavities and sac within which water may collect, the very porous concretious cortex, and the iron-rich plates toward the outside rusticle. Passageways (ducts) connect the central water column with the outside environment through these plates. Scales: left-hand diagram, 1 mm is equivalent to 20 mm; inset diagram, lower right, 1 mm is equivalent to 0.1 mm. (Figure originally published on page 102 of Ghosts of the Titanic by Charles Pellegrino 2000, published by HarperCollins Publishers Inc., New York.)

non when attempting the rehabilitation of a severely biofouled well in Ontario, Canada. Here, there was success in removing biological plugging that extended more than 15 m from the oxidative zone in the water well to very reductive zones well back in the formation material. In chemically analyzing the various water samples during posttreatment monitoring, it was noted that the various metallic elements appeared to have clustered along the gradient from oxidative to reductive. This gradient showed a grouping as follows (from oxidative to reductive with squared brackets indicating the grouping): [iron, copper], [zinc, manganese, titanium, chromium], [lead, barium, vanadium, cobalt], and [molybdenum, nickel, molybdenum], indicating, as a part of the bioaccumulation of some metallic cations, their location in the reduction–oxidation gradient. This location is dictated by the cationic species with some, such as iron, being at the very oxidative edge of the biomass whereas others such as nickel and molybdenum are at the reductive edge of the biomass. The consortial plugging occurring around this water well is therefore behaving like a biologically driven chromatograph.

Because the rusticles recovered from the *RMS Titanic* were growing in an oxidative environment, it is therefore not surprising that the dominant metal was iron, constituting 20%– 36% of the dry weight of the material. In the environmental industry, the absence of a particular metal does not mean that it is not present in that environment but simply that it was not detected in the sample. This finding does not exclude the possibility that all the cation in question has been bioaccumulated upstream from the position at which the sample was obtained. When undertaking environmental management of a porous medium that is biofouled, there remains always the possibility that hazardous chemicals could be bioaccumulated within the matrices of the biomass that could be released when there is a destabilization in that biomass. These accumulations could be viewed as the "icebergs" just over the horizon, as disasters that could one day happen without warning.

IV. Rusticle Relevance to Maritime Industries

The investigations to date of the rusticles have revealed that they are capable, under suitable conditions, of extracting iron from steel at significant rates. This biological extraction has the potential to seriously compromise the physical structure of a ship. The time frame for such compromise would appear to be based on the visual evidence. This evidence has been gathered from various sunken vessels (e.g., *Bismarck*, *Yorktown*, *Derbyshire*, the Japanese submarine I-52, and the *RMS Titanic*), located at various sites around the world. There remains the potential for the covert growth of rusticles within ships during the normal operational life span. Covert growth means that the rusticles could thrive at sites within the body of the vessel, particularly those that are not commonly inspected, and where conditions are conducive for growth.

In a ship's structure, areas that are most vulnerable include welded areas and zones which have severe stress concentrations (Mansour et al. 1997). It is estimated that two bulk tankers are lost every month, with 45% of these losses caused by heavy weather and structural damage. This category is further described as "strained crack in hull," concluding that structural failure is the major cause for the rising number of bulk carrier losses, as can be seen from the December 12, 1999, sinking of the *Erika* during a storm off the French coast. The 25-year-old tanker *Erika* broke in two, spilling about 13,600 t of fuel oil, polluting 400 km of beaches, and killing or maiming 300,000 seabirds. The classing agency for the tanker, Registro Italiano Navale Group (Internship classification & Management System Cert. Society), reported the initial findings into the cause of the accident as pointing to a small structural failure in hull structure. This structural failure led to further cracking and finally to the collapse of the hull (Hauley 2000). These structural failures can be a result of corrosion and fatigue cracking, which, in conjunction with biological attachment from within the tanker itself, can result in the loss of a ship's integrity (Ma et al. 1997). Such catastrophic failures of shipping have considerable long-term environmental pollution implications through the release of both degradable organics and the more recalcitrant inorganic materials, including toxic metals and radionuclides.

A range of factors would be important in considering the potential for these rusticles to grow rapidly enough to compromise the normal lifespan and seaworthiness of a ship (Cullimore and Johnston 2000b). These factors could include, but are not necessarily limited to, suitability of steel surfaces on which rusticles can form and function, a high level of humidity or a water-saturated environment, oxidative conditions, salt concentration in the water greater than 1.4%, temperature gradient, turbulence, nutrients, electrically charged surfaces, and neglect. A typical example of a condition where these rusticles could infest and compromise the integrity of the ship is between hulls and in compartments where a confined environment could provide conditions conducive to growth.

The most likely sites for a rusticle infestation to occur require a number of variables to be achieved, including surfaces or areas where the steel is poorly protected with paint, embrittled by stress, electrically charged in any way, involved in rhythmic movement of water over the site, positioned on a temperature gradient, or where available water contains sufficient nutrients to support growth. Where a site is not subjected to regular inspections, for example, on a monthly basis, or rusticle growth is not suppressed through the use of biocides or physical removal, the growths can then begin to extract iron from the steel and weaken the afflicted steel structures. It is a common practice to presume that the appearance of rusty encrustations are merely the result of physio-chemical activity and an inevitable part of the normal deterioration that may be expected. Traditionally, the appearance of rust within an enclosed chamber has not been viewed as a living mass that is "eating" away at the steel, but rather the rust is seen as an inevitable chemical event for which solutions may be ineffective over the long term.

In the water well industry, it is now acknowledged that most of the plugging and clogging events that occur down a well are actually biologically derived.

Comparable studies have revealed that the same groups of bacteria are involved in these events both down in water wells and deep down at the site of the *RMS Titanic*. Similar rusticle structures are observed at both sites. The question therefore becomes whether steel-fabricated ships floating on the surface, or the *RMS Titanic*, a splintered steel structure lying on the ocean's floor, are subject to the same bacterial challenges as water wells, which involve steel structures set into the groundwater. The arrival of nonindigenous organisms, such as the zebra mussel (*Dreissena polymorpha*) plaguing North American water systems, appears to have arisen from covert passengers in or on seagoing vessels. By analogy there may be "microbial" passengers traveling in the bilge areas and other damp dark places within ships that can cause problems to the ships themselves. In water wells, these nuisance bacteria can so substantially reduce water flow into a well that it has to be abandoned. The developing practice in the water well industry to prevent these infestations has evolved into the Sustainable Water Well Initiative in Canada. The objective is to extend the operating life span of wells by routine testing and suitable preventative maintenance or radical treatment procedures, depending on the state of the bacterial infestation. In the Canadian prairies, 200,000 operating wells have been found to have, on average, a life span of 15 yr. The capitalization of these wells has been estimated to be minimally CAN$1 billion, and the annual replacement and/or rehabilitation costs are CAN$67 million.

In the maritime industry, a similar scale of impact on the life span of ships also causes heavy replacement costs, pollution, or unscheduled repairs. These costs could be partially curtailed if the true level of these microbial impacts was found to be significant and controllable. In the shipping industry, there is also an average life span for ships, but sudden catastrophic sinkings continue to occur on a regular basis. By improving detection and control procedures to include microbial events that can weaken the structure of the ship, it should be possible to lengthen the life span of ships in much the same manner as water wells are becoming more sustainable.

V. Rusticle Biomass on the *RMS Titanic*

The examination of rusticle growth rates on the *RMS Titanic* has only been possible through annual or biennial expeditions to the wreck site. Through the use of video and high-resolution imagery, growth rates and patterns can be closely analyzed. In 1996, the first detailed examination of rusticles from the *RMS Titanic* was begun. This examination has permitted us to quantify rates and modes of rusticle growth on the outside of the bow section of the ship; however, the outside of the stern and the interior of the wreck cannot be clearly quantified because these areas remain largely unexplored. One aspect of this examination was to estimate the percentile coverage of the various parts of the bow section, together with the estimated thickness of the rusticles at these various sites. From the 1998 Expedition, comparative assessments through video imagery showed clear evidence that the rusticles continue to grow, and there is evidence that the biomass is approximately 30% greater than the mass observed in 1996.

Video surveys of the bow section of the *RMS Titanic* in August 1996 and 1998 allowed a quantitative estimate of the total volume of rusticles. This estimate is based on the area of the steel covered by rusticles at various points on the ship. Calculations reveal that the bow section of the ship had a total mass of 590 ± 30 t of rusticles in 1996, increasing to 800 ± 40 t in 1998. The iron content in the rusticles had presumably been extracted from the ship's steel structures and was now accumulated within the rusticles. The iron content has risen from 160 ± 10 t of iron in 1996 to 220 ± 12 t in 1998. Although a significant amount of iron is present within this mass of rusticles, there remains a concern as to the rate at which iron is being released from the rusticles into the oceanic environment. Iron concentrations vary in rusticles along with their size, weight, and density of the ducts on the rusticle surface. These estimates are based on examination of rusticle specimens recovered from both the 1996 and 1998 Titanic expeditions.

VI. Focused Accumulation Sites for Iron Within Rusticles

One concern arising from the examination of the *RMS Titanic* was the manner in which the iron is accumulating in the rusticles. It is known that the rusticles have a very large surface area and a highly porous concretious structure; however, little was known of the sites where the iron accumulates. Accumulation occurs as various forms of ferric oxide and hydroxide, dominated by goethites. To examine the sites of iron accumulation in the rusticles, a Veterinary Grade Radiograph machine MinXray 903 Type B-85, MinXray Inc., 3611 Commercial Ave., Northbrook, Illinois 60062-1822, operated at 100 kV, 1/20 sec, using Kodak high-speed film was used. Five rusticles, collected from the 1996 and 1998 expeditions, were investigated using radiographic examination (Cullimore and Johnston 2000a). It was found that the iron within the rusticle structure was not evenly distributed throughout (Fig. 2) but was concentrated into two major regions. The first region where iron was dispersed resembles "cloudlike" structures. The second region had very dense channeling of the iron into localized regions that spread weblike throughout the rusticle. The first impression was that the regions channeled with a high iron content bear resemblance to a primitive blood system.

The radiographic images confirmed the complex nature of the rusticle channels and the fact that the rusticles had entrapped small artifacts, coal fragments, and glass shards that billowed over the bow immediately after its collision with the ocean floor. Microbiological examination revealed that the microbial consortia within the rusticle were not evenly dispersed but remained concentrated within localized regions of the rusticle. No correlation could be established between the sites of iron concentration and the various bacterial consortia present in the rusticle.

It has been determined, based on extraction results, that a cycle has been

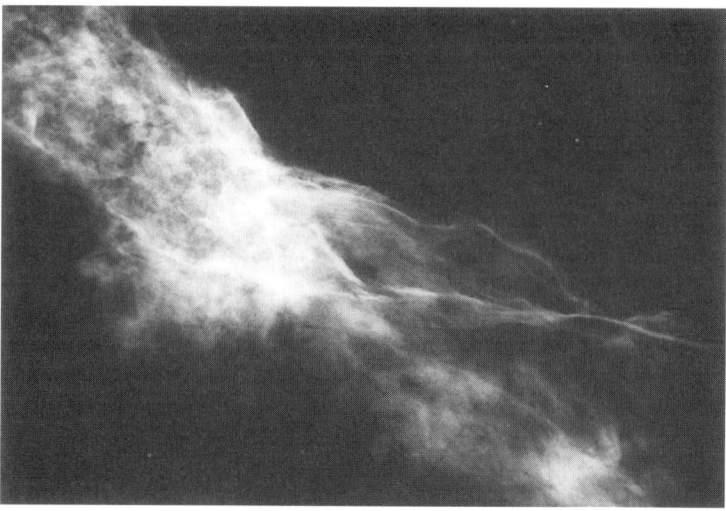

Fig. 2. Radiographic image of a rusticle recovered from the 1998 Titanic Expedition. This radiographic image represents one "shoulder" of a rusticle recovered from the bow section of the ship. The image reveals the density of the iron (the lighter the image, the greater the density of iron). The close-up image reveals there are complex structures within the rusticle body.

established in which iron is being biologically extracted from the steel of the ship into the rusticle structures. The iron is then exported into the oceanic environment as "red dust" (RD) and "yellow colloids" (YBC). The rate of extraction is increasing beyond the predicted 1996 rate of 0.09 t of iron being mined by the rusticles/d. This rate of dispersion of the iron through the release of RD and YBC into the greater oceanic environment is a critical factor in assessing the true nature of the environmental impact of the sinking of *RMS Titanic*. Thus, there is clearly a potential for some of the iron to enter into the surface biosphere as a result of events such as the consumption of fish and other animals from the oceanic environment. The iron in both the RD and YBC is likely to be consumed by the oceanic biota and, through that means, to enter the food chain. It is therefore possible that some of the iron from the *RMS Titanic* could actually find its way back into the blood of humankind and become a minor player in blood hemoglobin, primarily after entry into and passage through the phytoplankton biomass.

VII. Water Wells and Rusticles

A parallel can be drawn between the rusticles at the *RMS Titanic* and growths that occur within the various forms of wells that interface with groundwaters (Cullimore 1993; Cullimore and Johnston 2000a). The origin of the iron accumulating in these growths was, for the rusticles growing in the wells, mostly coming from the groundwater and surrounding geology. On the ship, the iron appears to be coming from the steel structure of the ship itself.

Downhole video inspection of water wells that are becoming biologically plugged reveals growths and concretions similar to those seen in the aforementioned caves and at the *RMS Titanic*. Injection and extraction wells at bioremediation sites reveal startlingly similar concretious growths. For example, a bioremediation site in Holland utilizes horizontal extraction wells as a part of the bioremediation process, extracting chlorinated hydrocarbons such as trichloroethylene and polycholoroethylene. The bioconcretious growths in the 10.16-cm well screen appeared within the first 2 yr of operation, causing a 50% loss in hydraulic production. These growths appear so morphologically similar to those observed at *RMS Titanic* that they may be called "sibling" species.

VIII. Laboratory Evaluation of Corrosive Processes

The first stage in the investigation has been to develop a methodology to determine whether microbial growths in the form of rusticles can cause iron extraction from a targeted steel plate. To achieve this outcome, a Biological Activity Reaction Test (BART™, Droycon Bioconcepts Inc., Canada; Cullimore and Alford 1990) was modified (Fig. 3). The BART™ test uses a 15-mL water sample

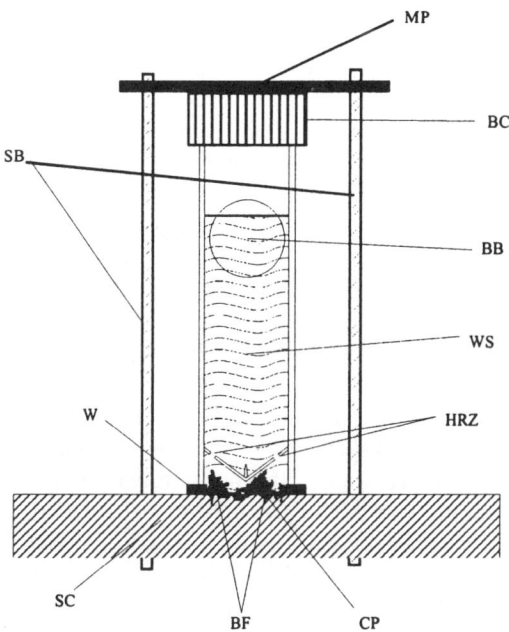

Fig. 3. The biological activity reaction test (BART™, Droycon Bioconcepts Inc., Canada) has been modified to determine corrosion of the surfaces of steel coupons.

that creates a reductive/oxidative (redox) front with a rising nutrient gradient selective to the bacteria being investigated (e.g., sulfate-reducing bacteria). To contact the mild steel coupon, six 3-mm holes were drilled through the basal cone of the test vial, and the vial itself was placed under pressure onto a 26-mm rubber washer to prevent leakage.

The BART test vial is modified to encourage the growth of bacteria within the water sample (WS) (see Fig. 3) on the steel coupon (SC). To achieve this, the BART test is placed on a circular rubber washer (W) and pressed down using a main plate (MP), which is screwed tightly on by two screw bolts (SB). The water sample is added before this procedure by unscrewing the BART cap (BC) and adding 15 mL of the water sample, at which time the BART ball (BB) floats up to create a redox gradient. Holes (HRZ) are cut in the base of the test vial to allow the bacteria to grow directly on the steel and form biofilms (BF). Corrosive pitting (CP) can be observed at the contact sites of the steel coupon (CP).

To conduct the test, a 1% (w/v) suspension of rusticles recovered from the site of the *RMS Titanic* was made using a 4% sterile solution of sea salts. The 15-mL suspension was aseptically injected through the cap into the test vial to set up a redox gradient selective for a specific group of bacteria. The tests were conducted at room temperature (21 °C) for 60 d. Each steel coupon was examined for biological corrosion or extraction by reflectance microscopy (×10 and ×40). It was observed that biological growths appear to be attached to the steel surfaces on which the test vial had been positioned. Examination of the growths by both photomicrography and radiography revealed a range of structures that resembled rusticles in color, texture, and form. The observed rusticle-like structures grew from the surface of the mild steel to an average height of 2 mm, occupying 60% of the volume (after 60 d), having a wet volume of approximately 0.37 mL, a dry weight of 0.09 g with an iron content of 22%; this resulted in approximately 0.2 g of iron being biologically extracted from the steel in the 60-d period. Removal of the rusticle structures attached to the steel and subsequent cleaning of the steel was achieved using a chemical immersion technique. The steel coupon was immersed in a 1.0% solution of hydrochloric acid for 60 min, followed by a sterile distilled water rinse. The steel was again microscopically examined for evidence of biocorrosion and/or extraction. Pitting was observed covering 20% of the steel's surface, with the pits ranging in diameter from 0.1 to 0.8 mm with a depth of 0.4–1.0 mm.

This experiment revealed that, under reductive conditions, the biocorrosive processes could be initiated within 60 d at room temperature. These processes generated biological growths that resembled rusticles on a microscale. The next stage of this study will be to apply this methodology to coupons of steel from the *RMS Titanic*. This experiment will provide somewhat controlled conditions to observe the forms of corrosion or extraction that occur on this particular steel. Field experience has revealed that corrosion or extraction tends to be dominated by lateral flaking and dissolution of the steel directly under maturing rusticle growth.

IX. Environmental Cost of Covert Rusticle Growth

There is a clear chain of evidence that suggests microbial growths such as rusticles, in the form of slimes, tubercles, nodules, and encrustations, can extract elements such as iron from the steel into these structures. This type of functioning bioconcretion is evident at the *Titanic* site. These growths are contributing to the deterioration of the *RMS Titanic* through biodegradation; it is a natural extension to consider the implications of these biofunctions at other sites where conditions may be conducive for such growth. It becomes clear that the rusticles have become dominant in an environment where there is high salt concentration (i.e., highly conductive), available organic material such as the "sea snow" found at the wreck site, and steel surfaces.

In examining existing engineered steel structures present within the oceanic environment, it is evident that current management strategies, over the lifetime of such structures, would be ineffective and restrictive to the determination of observable occurrences of rusticle-like growths. There are two main points at which these growths may become key to the ongoing life of a structure. The first critical point is that these growths may commonly be generated at covert sites, remaining unobserved, yet still remain able to create catastrophic failure of the structure. Loss of buoyancy, leaking, and total loss of structural integrity could follow, resulting in the sinking of the vessel and rendering forensic proof for this type of failure difficult to determine. The second critical point relates to abandoned steel-fabricated structures, such as pipelines, oil rigs, drums of ocean-dumped hazardous waste, reactor casings of lost atomic submarines, and plutonium triggers aboard these submarines. Rusticle colonies now have the potential to seriously compromise the structural integrity and containment of these artifacts. To a lesser degree, rain- and salt-exposed crevices in the supporting structures of bridges may be vulnerable to rusticle activity.

Historically, there has been very little understanding or examination of the impact of microbial activity on metallic structures. The extent and significance of these biologically induced events must rely heavily on the limited knowledge that has been gained with respect to biocorrosion and on the understanding of the function and aggressivity of microbial communities (consortia) to establish and proliferate in extreme environments. The covert microbial growth and degradation of steel structures has the potential to cause economically significant environmental problems.

Potential maritime examples range widely, but perhaps one that has a long-term and serious concern is the disposal of steel drums containing toxic or nuclear wastes into the marine environment over the past 50 yr. The chain of events established by this practice involves the loading of an industrial steel drum with radioactive material and allowing it to free fall to the ocean floor. At the time this practice was established, popular opinion was that the ocean floor was a sterile environment divorced from the terrestrial environment. In the event of a free fall of the loaded drum, the first concern is generated by the collision between the drum and the ocean floor, as well as high-pressure implosion of air pockets contained within the drum itself. These impacts may cause seam failures or the embrittlement of the mild steel of the drum. These sites then become focal points for initial attachment and growth of microorganisms. The form and extent of the growth would depend on the nutrients in the environment (positive impact) and the level of radiation being generated by the wastes (negative impact).

In the case of a negative impact caused by radiation, two microbial mechanisms can counter this influence on their growth. The first of these mechanisms is that some microorganisms, such as the bacteria belonging to the genus *Deinococcus*, have a very high resistance to radiation (e.g., 1–2 Mrad) due to unique cell-wall-surface structures and DNA-repair systems. Under suitable environmental conditions, these bacteria could therefore grow over the surface of the steel containment drum, causing a loss of integrity. An alternative mechanism occurs where the indigenous microbial consortia are protected from radiation by the nature of the concretious growths that these microbes form. The microbially induced concretious growths contain a high metal content, retarding the impact of the radioactive contents of the drum. Initial growth may occur through the metal-rich biocolloids, such as red dust from the growth of rusticles, which attach or come to rest on the outer surfaces of the steel structure. Consequent growth may then generate conditions that cause corrosion of the steel structure, followed by perforation and leakage of the contained radioactive waste into the environment. The biological embarrassment of steel containment structures such as these rums is likely to occur over an extended period of time, yet the potential for large-scale environmental contamination continues to be significant and largely unpredictable.

The potential for corrosion by rusticle-like concretions on drums filled with toxic material is probable and problematic. This concept is founded on the dual premise that (1) there is no restricting radiation field to suppress microbial activity, and (2) the contained toxic material may well be selectively toxic to humankind and the biota but not toxic to many species in the microbial community. For example, cyanide is clearly perceived to be a powerful toxic compound; however, many microorganisms are resistant to cyanide and some can even degrade it as part of the process of natural self-purification commonly seen when toxic spills occur. The sequences involved in the event of the degradation of toxic containment vessels would follow a pattern of impact, embrittlement, localized corrosion, and, finally, perforation. Perforation would be followed by localized leakage, with a probable accumulation of the toxic material within the concretious growths. Bioaccumulation events would occur at the viable cells present and active within the concretious growth. Finally, biodegradation or dispersion, depending on the nature of the toxic material, would occur. Growths such as rusticles would therefore both accelerate, through corrosion and perforation of the steel structure, and reduce, through bioaccumulation with or without biodegradation, the release of toxic materials from the compromised structure.

Underwater pipelines offer a number of problems relating to the potential biological compromise that could involve the activity of rusticles. The contents of the pipeline, whether it is abandoned or in use, may be critical to the nature of the growth. For example, an underwater gas pipeline may suffer from corrosive physical pitting or perforation, leaking methane into the surrounding environment where it would be utilized by the methane-consuming (methanotrophic) bacteria (MCB). These MCB may be incorporated into the concretious consortium-forming rusticles. Consequently, early microleaks of gas or oil may serve as the primary organic feedstock for the rusticles, which then become much more aggressive and, in so doing, structurally weaken the pipe, causing sudden and dramatic pipeline failures. Abandoned pipelines would carry only static perched material that may form a smaller feedstock for microbial growth, but the lack of forced flow and pressures may allow the microorganisms to gain entry into the pipeline itself and initiate reductive corrosive processes from directly within the pipeline. Concurrently, the pipeline could also be challenged with oxidative corrosive processes being generated by rusticle-like structures. The form of these attacks on the integrity of the steel pipeline would be very much affected by the nature of the variable electrical charges along the pipeline because anodically charged sites tend to be the focal point of the microbial activity. Many pipelines are protected, during their active life span, from these types of events through a process of cathodically charging the outer surfaces, which reduces microbial attachment and growth. Once the pipeline is abandoned, corrosion prevention practices are halted, and the gradual process of microbially induced corrosion and deterioration continues unimpeded.

In offshore drilling operations such as those for oil and gas, the steel structures being used for such operations are known to be subjected to corrosive processes generally thought to have originated from infestations of sulfate-reducing bacteria (SRB). The presence of this type of activity is clearly signaled by hydrogen sulfide-driven electrolytic corrosion, the presence of copious black-sulfide-rich slimes, and, often, the "rotten egg" odor of the gas. This group of microorganisms forms a major component of the consortia of rusticles and can be found cloistered within the bioconcretions formed by this process. Rusticle growths on the outside of a drilling rig would clearly be recognized, treated accordingly, and removed during routine maintenance procedures. The same thing may not happen when rusticles grow within the ballast tanks of these rigs. Where this process is occurring, there is a risk that the growth may compromise the structural integrity of these ballast tanks in a manner that could lead to sudden and catastrophic failures.

Abandoned steel rig structures, particularly when immersed in seawater beneath the light penetration zone, are more likely to become infested with bioconcretious structures resembling rusticles. Residual hydrocarbons would form a major part of the feedstock supporting the growth, and the primary focal regions for growth are likely to be oxidative over steel that is, in some manner, embrittled to facilitate microbial attachment and growth. The rate at which these rusticle-like structures would begin to form and compromise the steel cannot be estimated because the experiential base is largely limited to a few sunken ships, with the *RMS Titanic* being the one most vigorously investigated. Ships sunk in shallower depths, such as the *RMS Britannic*, show very different forms of biological encrustations with a much greater diversity of life forms. The relative significance of the relationship between these different species and the rate of corrosive challenges to the steels on the vessels remains to be determined.

The potential environmental costs of biological activities on submersed steel structures are a combination of positive and negative impacts. Positive impacts relate to the recycling of the iron from the structures back into the natural oceanic cycles. This impact relates, in part, to the *RMS Titanic* investigation, forming perhaps a fitting conclusion to the investigation of the ship herself. Of particular interest is the fact that iron has now become recognized as a key limiting factor in the rate of oceanic carbon dioxide fixation through the phytoplankton. The iron released from these submerged steel structures may therefore play a significant role in increasing the ability of the oceans to fixate carbon dioxide and therefore, in some small way, thus reduce the potential impact of increased carbon dioxide on global warming. From there, the iron would gradually move into the global biosphere so perhaps, some time in the distant future, there will be at least one or two iron atoms from the *RMS Titanic* pulsing through every human body. The same may be said to occur in the fullness of time for the many other steel ships now resting on the ocean's floor and gradually deteriorating. This result may be viewed as a positive long-term environmental effect.

Some negative environmental impacts are more diverse and difficult to predict. These impacts may relate more to the releases of toxic, potentially carcinogenic, or recalcitrant chemicals of concern that were previously confined by steel but are now being released, at least in part, as a result of biological activity. These effects become very difficult to predict, challenging to contain, and may impact in many different manners over various time scales. In the past, when a ship sank at sea, there was little concern for its impact on the environment. Today, a ship sinking means simply that the contents of that ship have now become subject to a different pattern of release that is not in the domain of humankind. Impacts, therefore, become speculative because the knowledge has yet to be gained, and in that regard the *RMS Titanic* is one of the signposts pointing the way to understanding these effects.

There is a continuing need for dedicated research and development to address the capacity of rusticles and other microbial communities to cause sudden losses in steel strength and integrity. These losses must be recognized and effectively managed throughout the life span of the steel structure, but also after usage. This postusage monitoring will, in effect, require a more complete understanding of biologically induced corrosion events and how these activities are impacting the environment.

Three different replicable consortially driven forms of growth are proposed to form the foundation of a new scheme for the recognition of living entities formed and dynamically driven by a multiplicity of microbial species that may commonly contain bacteria and fungi. Each of these entities would have a generic name that would relate to a commonality of form and consortial composition, whereas the species name would reflect the habitat where the entity was most commonly observed growing. The following represent suggestions within the mandate just discussed. Rusticles are defined as the "type" consortium.

The other two consortially formed living entities discussed here relate to the black plug layering in golf courses and the plugging of water wells. In these events, the consortia possess two common characteristics. First, the growth is formed by the glycocalyx to produce a coherent body of "slime." Second, there is an occlusion or loss in permeability (plugging) in the infested porous medium. For the black plug layer, this can be a dramatic event, with permeability falling several orders of magnitude over a matter of weeks. This change is caused by a rapid shift in the voids from being essentially saturated with water to a state of being totally occupied by the "slime" (glycocalyx) generated by the consortium. For the plugging, the loss in hydraulic conductivity can be a very gradual happening as the biofilms forming the plug (glycocalyx) only slowly fill the void space. There are thus two major consortial groups proposed under this heading: (1) the black plug layering consortium and (2) the plug-forming bacteria that commonly accumulate iron along with other metals within their growths. As both these forms of growth occur within porous media, their form and structure become more difficult to determine because direct observation is not possible except under laboratory conditions.

In these consortia, there is a form of social "intelligence" that is one step beyond "quorum sensing," for here the various incumbent species through cooperation (assigning everything to its proper proportion) generate a concord within which all incumbents can mutually benefit. Although this might not be true of the bulk of plants and animals, it may be reasoned that most microorganisms forming a part of the larger environment do form into consortia, which may vary from tenuous to essential and commonly are within an extracellular matrix of some type. In understanding each of these component species, their role within the consortia needs to be understood as an integrated part of the social "intelligence" that is inherent in the consortium. We clearly have come to accept the defining role of species in the classification of living organisms. Perhaps, when examining the microbial kingdom, classification should now be based on the living entities generated by consortial synthesis and maturation rather than on a study of those components that can, through convenience, be readily observed using traditional scientific concepts.

Perhaps the most challenging and fascinating aspect of the presence of these often covert consortia is their role in the evolution of life forms on this planet. Present concepts cloister species as unique entities capable of independent existence. Given the extreme environments prevailing during the early development of the biosphere on planet Earth, it would not be unexpected to find that the simple life forms evolving at that time would either have found specialized environments such as high salinity, extreme heat, or radical pH, or developed some mechanisms for a compensatory synergistic development with other species. Today, the Archeaobacteria remain dominant in the extreme environments. In the surface biosphere, we have yet to determine the full extent and nature of the microbial synergy that would have led to the evolution of complex and robust consortia such as the ones now on the *RMS Titanic*, in golf course greens, and bioplugging water wells.

It is not unreasonable to conjecture that these consortial forms may be considered as living fossils dating back 3 billion yr to the inception of multicellular life and represent not only definable life forms but the beginnings of tissue differentiation. Of all the consortia so far discussed and recognized, the rusticles display the most differentiation. Through the cooperative activities of a range of Eubacteria, Archaeobacteria, and Fungi, the rusticle shows differentiable structural forms and activities that would be more expected of an animal than of a collective group of protists. The central water channel, reservoirs, passageways, and ducts to the outside possessed by the rusticles indicate a rudimentary circulatory system. In the laboratory, the venting of gases (principally CO_2) would imply that the gases could act as a pneumatic mechanism for moving water through the rusticle. The large plates and columns rich in ferric iron perhaps represent a primitive skeletal structure providing physical integrity to the consortial matrix. The very porous nature of the consortium, with a high sorption capacity, would allow the retention of nutrients, metals, and both anions and cations in forms that would augment the ongoing growth of the rusticle. The metal in the hull of the *RMS Titanic* has come to form essentially a culture substrate for the growth of rusticles.

It would therefore be of considerable scientific interest to determine in a more precise manner the roles that microbial consortia may well have played in the evolution of the animal and plant kingdoms in an open-minded analytical manner. Marcus Aurelius (1969b) considered the need to address the dogma that sometimes impedes progress by observing:

> The Pythagoreans bid us in the morning to look to the heavens that we might be reminded of those bodies which continually do the same thing and in the same manner perform their work, and also be reminded of their purity and nudity. For there is no veil over a star.

In our dogmatic application of the classification of living systems using the species concept, we have perhaps pulled a "veil" over the understanding of the dynamics of living entities that do, in the form of consortia, flourish.

X. Conclusion

There is little doubt that the *RMS Titanic* was traveling close to her maximum speed as she entered that fatal field of ice. Time was dictating speed, and arrogance spiced with greed ignored the ice, but there was a collision with the ice and the ship sank quickly with a tragic loss of life. Today the "unsinkable" ship lies broken on the ocean floor, but not at rest, for Nature is now recycling the iron, transforming the ship, and folding it back into the web of life. The time is now to learn the lesson woven by Nature into the ship as the ship yields back her being to become only a distant memory.

The lessons relevant to environmental pollution and toxicology relate to the nature of the microbial events that are now moving the iron and other elements away from the hull into the local oceanic environment. Perhaps the most significant is that a complex of microbial consortia have, in a very obvious manner, created rusticles throughout the oxidative environments both outside and within the hull. These consortia have collectively generated bioconcretious-like materials within and upon which the consortia operate. A major net outcome of this is the export of iron from the steel via the rusticles into the oceanic environment as red dust and yellow biocolloids. Within these particles are contained the biologically extracted iron and other metallic cations that are primarily bioaccumulated within the complex of polymeric structures which surround the microbial cells forming the consortium. Had the rusticles not been so obvious, then the condition would have been very similar to the biofouling that occurs in and around water wells where it is not possible to view these rusticle-like structures formed by the consortia.

In bioremediation work, there is still the mindset that every function has to be performed by a single species and not by a consortium. The outcome of this lateral thinking from the diagnosis of acute diseases is that most biologically driven events are thought to be a direct result of the activity of a single species of microbe. As a result, much effort is presently being devoted to the molecular-level determination of very particular species that are considered to be totally responsible for the problem or solution being managed. No consideration is given to the potential for one or more consortia to be effectively driving the problem or the solution. In the next two or three decades, the prime lesson from the environmental impacts of the *RMS Titanic* on the oceanic environment will be that the biological activities observed have been totally driven by microbial consortia. In these coming decades, the role of these sometimes covert consortia impacting in many environments, including the human body, will need to be addressed and recognized if a more mature management of environmental resources and remediation is to be achieved. Today, the frontiers are moving away from the *RMS Titanic* to the plugging of oil wells, the curing of concrete, the nature and diagnosis of chronic diseases, the rehabilitation of plugged water wells, and the extreme adaptability of these microbial consortia to extreme environments both on this planet and elsewhere.

Summary

The *RMS Titanic* sank in 1912 and created a historical event that still ripples through time. Stories were told and lessons learned but the science has only just begun. Today the fading remains of the ship resemble the hanging gardens of Babylon except that it is not plants that drape the walls but complex microbial growths called rusticles. These organisms have been found to be not a species, like plants and animals, but to be structures created by complex communities of bacterial species. Like the discovery of tube worms in the mid-oceanic vents, the nature of these rusticles presents another biological discovery of a fundamental nature. Essentially these microbial consortia on the *RMS Titanic* have generated structures of a mass that would rival whales and elephants while gradually extracting the iron from the steel. Rusticle-like consortia appear to play many roles within the environment, and it is perhaps the *RMS Titanic* that is showing that there is a new way to understand the form, function, and nature of microorganisms. This understanding would develop by considering the bacteria not as individual species functioning independently but as consortia of species functioning in community structures within a common habitat. This concept, if adopted, would change dramatically the manner in which a microbial ecologist and any scientist or engineer would view the occurrence of a slime, encrustation, biocolloid, rust flake, iron pan, salt deposit, and perhaps even some of the diseases that remain unexplained as a disease of unknown cause.

Acknowledgments

The authors acknowledge the financial support of the Society of Naval Architects and Marine Engineers for this project. Also acknowledged are the many people who at different times provided important information or facilitated the collection and analysis of the rusticles: these include Bill Garzke (Gibbs & Cox Ltd.), George Tulloch, Greg Andorfer (Maryland Science Center), P.H. Nargeolet (Infremer), David Elisco (Stardust Visual), Charlie Haas, and the staff of Droycon Bioconcepts Inc.

References

Alford GA, Cullimore DR, Monea M, Ostryzniuk N (2001) Device for the Blended Cathodic Disruption of Biofouled Sites. U.S. Patent Pending.

Ballard R (1987) The Discovery of the Titanic. Warner Books, New York.

Cullimore DR (1993) Practical Manual of Groundwater Microbiology. CRC Press, Boca Raton, FL.

Cullimore DR (1999) IV Titanic: the connection between rusticles and clogging. In: Alford, GA. Cullimore R (eds) The Application of Heat and Chemicals in the Control of Biofouling Events in Wells. Lewis, Boca Raton, FL.

Cullimore DR (2000a) Microbiology of Well Biofouling. Lewis, Boca Raton, FL.

Cullimore DR (2000b) Consortial bacterial forms. In: Practical Atlas for Bacterial Classification. Lewis, Boca Raton, FL, pg. 131–135.

Cullimore DR, Alford GA (1990) Method and apparatus producing analytic culture. U.S. Patent 4,906,566.

Cullimore DR, Johnston L (2000a) The impact of bioconcretious structures (rusticles) on the *RMS Titanic*: implications to maritime steel structures. In: Proceedings of the Annual Meeting of The Society of Naval Architects and Marine Engineers, Vancouver, British Columbia, October 2000, paper 9.

Cullimore DR, Johnston L (2000b) Biodeterioration of the *RMS Titanic*. Canadian Chemical News, November/December. The Chemical Institute of Canada, Ottawa, Canada.

Davis DG (1989) Helictite bushes, a subaqueous speleothem? Natl Speleol Soc Bull, 51: 120–124.

Davis DG, Palmer MV, Palmer AN (1990) Extraordinary subaqueous speleothems in Lechuguilla Cave, New Mexico. Natl Speleol Soc Bull, 52:70 B86.

Droycon Bioconcepts Incorporated (1999) BART™ Biological activity reaction test user manual. Droycon, Saskatchewan.

Ewald PW (2000) Plaque Time: How Stealth Infections Cause Cancers, Heart Disease, and Other Deadly Ailments. Simon and Schuster, New York.

Garzke WH, Brown DK, Matthias PK, Cullimore R, Wood D, Livingstone D, Leighley HP, Foecke T, Sandiford A (1997) *Titanic*, The Anatomy of a Disaster. Report from the marine forensic panel (SD-7). Proceedings of the 1997 Annual Meeting of the Society of Naval Architects and Marine Engineers, Ottawa, Canada.

Gerhules J, Alford A (1990) Application of physico-chemical treatment techniques to a severely biofouled community well in Ontario, Canada. In: Howsam P (ed) Water Wells Monitoring, Management, Rehabilitation. Spon, London.

The Harvard Classics edited by Charles W. Eliot, LL.D.1 P. F. Collier & Son Corporation, New York, 1969.

Hauley B (2000) Safety charter targets old tankers, flags of convenience. Maritime Wkly Maritime Report Eng News (March).

LaRock EJ, Cunningham KI (1995) Helictite bush formation and aquifer cooling in Wind Cave National Park, South Dakota. Natl Speleol Soc Bull, 57:43–51.

Ma K-T, Orisamolu IR, Bea RG, Huang RT (1997) Towards optimal inspection strategies for fatigue and corrosion damage. In: Proceedings of the 1997 Annual Meeting of the Society of Naval Architects and Marine Engineers, Ottawa, Canada, SD-7.

Mann H (1997) A close-up look at Titanic's rusticles. Voyage 25 47–48 pp.

Mansour AE, Wirsching PH, Lucket MD, Plumpton AM, Lin YH (1997) Structural safety of ships. In: Proceedings of the 1997 Annual Meeting of the Society of Naval Architects and Marine Engineers, Ottawa, Canada, SD-7.

Marcus Aurelius (1969a) The Meditations of Marcus Aurelius (trans. G Long) Book V, observation 30, pg. 230. The Harvard Classics edited by Charles W. Eliot, LL.D. P. F. Collier & Son Corporation, New York, 1969.

Marcus Aurelius (1969b) The Meditations of Marcus Aurelius (trans. G Long) Book XI, observation 27, pg. 293. The Harvard Classics edited by Charles W. Eliot, LL.D. P. F. Collier & Son Corporation, New York, 1969.

Pellegrino C (2000) Ghosts of the Titanic. Harper Morrow, New York.

Pellegrino C, Cullimore DR (1997) A study of the bioarcheology of a physically disrupted sunken vessel. Voyage 25 39–46 pp.

Sly LI (1995) Microorganisms, an essential component of biological diversity. Can Soc Microbiol Newsl 42(2):7–10.

Straus E (1997) Mob action, peer pressure in the bacterial world. Sci News 152:124–125.

Wells W, Mann H (1997) Microbiology and formation of rusticles from the *RMS Titanic*. Resour Environ Biotechnol 1:271–281.

Manuscript received May 1, accepted May 11, 2001.

Index

Acetylcholinesterase inhibitors, tissue residues, 17
AChE inhibitors, CBRs, 5
AChE inhibitors, mode of action, 8
AChE inhibitors, tissue residues, 17
Acridine & its homocyclic analog anthracene (illus.), 41
Acridine, anaerobic degradation pathway, 45
Acridine, photoenhanced toxicity different species (table), 60
Acridine toxicity, species groups, 50
Ah-receptor agonists, CBRs, 5
Ah-receptor agonists, defined, 2
Ah-receptor agonists, mode of action, 8
Ah-receptor agonists, tissue residues, 20
Anaerobic degradation pathway, acridine, 45
Aquatic organisms, CBRs, 2
Aquatic organisms, contaminant residues, 1 ff.
Aquatic organisms, exposure vs toxicological effects, 4
AQUIRE database, website address, 48
Archeaobacteria, dominant in extreme environments, 137
Atropisomer of methylsulfonyl-PCBs, 91
Atropisomers of PCBs, chiral PCBs, 91
Azaarene toxicity, 39 ff.
Azaarenes, biotransformation and toxicity, 52
Azaarenes, carcinogenicity, 62
Azaarenes, chronic effect different species (graph), 54
Azaarenes, chronic toxicity, 52
Azaarenes, defined, 40, 42
Azaarenes, direct toxicity, 48
Azaarenes, genotoxicity, 62
Azaarenes, K_{ow} related to baseline toxicity, 48
Azaarenes, metabolites in the environment, 47
Azaarenes, photo-products affect photosynthesis of marine diatom, 61

Azaarenes, photochemical reactions, mechanisms & kinetics, 53
Azaarenes, photochemical transformation, 53
Azaarenes, photoenhanced toxicity, 59
Azaarenes, phototoxic effects, 53
Azaarenes, risk assessment, 69
Azaarenes, teratogenicity, 66
Azaarenes, toxic effects several to different species (graph), 51
Azaarenes vs homocyclic PAHs, comparative toxicity, 67
Azaarenes with more than three aromatic rings, microbial degradation, 46

BART™ (Biological Activity Reaction Tests), 124, 131
Benzacridines, mutagenicity, 65
Benzenehexachloride (BHC)(HCH), world emission, 93
Benzoquinoline metabolism (illus.), 47
Benzoquinolines & metabolites, mutagenicity, 64, 65
Benzoquinolines, genotoxicity, 63
Benzoquinolines, metabolic routes, 44
BHC (HCH), world emission, 93
Biological Activity Reaction Tests (BART™), 124, 131
Black plug layering consortium (microbial), 137
Boranes, polychlorinated, chiral insecticide, 91
Bromocyclen, chiral insecticide, 91

Carassius auratus, chlorinated tissue residues, 12
Carbazole & homocyclic analog fluorene (illus.), 41
Carcinogenicity, azaarenes, 62
CAS numbers, chiral pesticides, 107
CBR approach, uncertainties, 6
CBR defined, 2
CBR determinants, 10
CBR evaluation, methodology, 6

143

CBR variability, 9
CBRs (Critical Body Residues), 2
CBRs, acute/chronic toxicity of chemical classes, aquatic organisms, 5
CBRs, ranges affecting survival, 27
Chemical classes, modes of action, 8
Chiral compounds, enrichment processes, 105
Chiral compounds, ERs in different compartments, 93
Chiral compounds, mirror image structures, 92
Chiral compounds, shielding from the racemate, 104
Chiral PCBs, 91
Chiral pesticides, constant enantiomer fraction, 101
Chiral pesticides, enantiomeric enrichment in environment, 85 ff.
Chiral pesticides, enrichment processes, 105
Chiral pesticides, list, 90
Chiral pesticides, racemic mixture deviations, 100
Chiral pesticides, shielding from the racemate, 104
Chiral pesticides, worldwide use, 86
Chlordane compounds, chiral insecticides, 90, 94
Chlordane compounds, stereochemical recognition, 103
Chlordane-*cis*, chiral insecticide, 90, 97, 104
Chlordane-*trans*, chiral insecticide, 90, 98, 104
cis-Chlordane, CAS number, 107
cis-Chlordane, chiral insecticide, 90, 97, 104
cis-Chlordane, enantiomer structures, 92
cis-Chlordane, IUPAC chemical name, 107
CNS seizure agents, CBRs, 5
CNS seizure agents, defined, 2
CNS seizure agents, mode of action, 8
CNS seizure agents, tissue residues, 19
Comparative toxicity, azaarenes vs homocyclic PAHs, 67
Consortial microbial events, deep sea rusticles, 117 ff.
Consortial nature of microbial events, *Titanic*, 117 ff.
Contaminant effects, aquatic organisms, 1 ff.
Critical Body Residues (CBRs), 2

D*aphnia magna*, narcotic tissue residues, 11
2,4'-DDD, chiral insecticide, 91
2,4'-DDT, chiral insecticide, 91
DDT residues, mummichogs, 21
Deinococcus spp., radiation-resistant bacteria, 134
Dibenzacridines, mutagenicity, 66
Dichlorprop, chiral herbicide, 90

EF (enantiomer fractions), defined, 88
Enantiomer fractions (EF), defined, 88
Enantiomer fractions (EFs), chiral compounds, 102
Enantiomeric enrichment, chiral pesticides in environment, 85 ff.
Enantiomeric enrichment, methodology, 88
Enantiomeric ratio (ER), defined, 86, 88
ER (enantiomeric ratio), defined, 86, 88
ER, as a tracer tool in environmental studies, 87
Excitatory agents, CBRs, 5
Excitatory agents, mode of action, 8
Excitatory agents, tissue residues, 15

Fenvalerate residues, rainbow trout, 21

Genotoxicity, azaarenes, 62
Glycocalyx, "slime", 137

HCH (technical), world emission, 93
α-HCH (hexachlorocyclohexane), chiral insecticide, 90, 93, 95
α-HCH, enantiomer structures, 92
γ-HCH (lindane), chiral insecticide, 93
HCH, CAS numbers, 107
HCH, IUPAC chemical name, 107
HCH-α (hexachlorocyclohexane), chiral insecticide, 90, 93, 95
HCH-γ (hexachlorocyclohexane) lindane, chiral insecticide, 93
Heavy metals, mode of action, 8

Heptachlor *exo*-epoxide, CAS number 107
Heptachlor *exo*-epoxide, chiral insecticide, 90, 99, 104
Heptachlor *exo*-epoxide, enantiomer structures, 92
Heptachlor *exo*-epoxide, IUPAC chemical name, 107
Hexachlorocyclohexane (α-HCH), chiral insecticide, 90, 93, 95
Hexachlorocyclohexane (HCH), nine stereoisomers, 93
Hexachlorocyclohexane (HCH), seven mesoforms, 93
HOMO-LUMO gap energies, PAHs, 59

Inorganic metals, tissue residues, 23
Isoquinoline, microbial degradation pathway, 44
Isoquinoline, toxicity, 42
IUPAC chemical names, chiral pesticides, 107

K_{ow}, azaarenes related to baseline toxicity, 48

Lindane (γ-HCH), chiral insecticide, 93

Mecoprop, CAS number, 107
Mecoprop, chiral herbicide, 90, 94, 96
Mecoprop, enantiomer structures, 92
Mecoprop, IUPAC chemical name, 107
Methane-consuming bacteria, rusticle component, 135
Methodology, enantiomeric enrichment, 88
Microbial consortial events, deep sea rusticles, 117 ff.
Microbial consortium, defined, 122
Microbial degradation, azaarenes more than three aromatic rings, 46
Microbial degradation pathway, isoquinoline, 44
Microbial degradation pathway, quinoline, 43
Mirror image structures, chiral compounds, 92
Modes of action, chemical classes, 5, 8
Modes of action, narcotic chemicals, 3, 5

Mutagenicity, benzacridines, 65
Mutagenicity, benzoquinolines & metabolites, 64, 65
Mutagenicity, dibenzacridines, 66
Mutagenicity, quinoline, 63
Mutatox™, azaarene genotoxicity tests, 62, 64

N-heterocyclic PAHs, 40
Narcotic chemicals, 7 ff.
Narcotic chemicals, modes of action, 3
Narcotic chemicals, polar/nonpolar, 3
Narcotics, CBRs, 5
Narcotics, mode of action, 8
Nuclear waste, steel drum deterioration on ocean floor, 134

Organometallic chemicals, tissue residues, 25
Organometals, mode of action, 8
Oxychlordane, CAS number, 107
Oxychlordane, chiral insecticide, 90, 100, 104
Oxychlordane, enantiomer structures, 92
Oxychlordane, IUPAC chemical name, 107

PAH, chemical reaction in water column (diagram), 55
PAHs (polycyclic aromatic hydrocarbons), 40
PAHs, biotransformation and metabolism, 41
PAHs, HOMO-LUMO gap energies, 59
PAHs, phototoxic effects on aquatic organisms, 57
PCBs, atropisomer of methylsulfonyl-PCBs, chiral PCBs, 91
PCBs, chiral, 91
PCBs, tissue residues aquatic organisms, 22
PCDDs, tissue residues aquatic organisms, 22
PCDFs, tissue residues aquatic organisms, 22
β-pentachlorocyclohexene, chiral PCB, 91
γ-pentachlorocyclohexene, chiral PCB, 91
Pentachlorocyclohexene-β, chiral PCB, 91
Pentachlorocyclohexene-γ, chiral PCB, 91

Pesticides, chiral enrichment processes, 105
Pesticides, chiral list, 90
Pesticides, pure enantiomer advantages of using, 86
Pesticides, racemic disadvantages of using, 86
Phenanthridine, biotransformation products (illus.), 46
Phenoxypropanoic acids, chiral herbicides, 90
Photoenhanced toxicity, aquatic contaminants, 13
Photoenhanced toxicity, azaarenes, 59
Photosynthesis, azaarene photo-products effects on diatom, 61
Phototoxic effects, azaarenes, 53
Phototoxic effects on aquatic organisms, PAHs, 57
Pimephales promelas, pentachlorophenol tissue residues, 16
Plug-forming bacteria, iron accumulating, 137
Polychlorinated boranes, chiral insecticide, 91
Polycyclic aromatic hydrocarbons (PAHs), 40
Pure enantiomer pesticides, advantages of using, 86

QSARs (Quantitative Structure-Activity Relationships), 1 ff.
Quantitative Structure-Activity Relationships, (QSARs), 1 ff.
Quinoline, microbial degradation pathway, 43
Quinoline mutagenicity, 63
Quinoline toxicity, 42
Quinoline toxicity, species groups, 49

Racemic pesticides, disadvantages of using, 86
Reactive chemicals, tissue residues aquatic organisms, 18
Reactives/inhibitors, CBRs, 5
Reactives/inhibitors, mode of action, 8
RMS Titanic, see Titanic, 117
Rusticle, environmental cost of covert growth, 133

Rusticle, radiographic image (illus.), 130
Rusticle, typical components of dissected rusticle (illus.), 125
Rusticle, word derivation, 123
Rusticles, chemical composition, 124
Rusticles, consortial microbial structures, 121
Rusticles, focused accumulation sites for iron, 129
Rusticles, found on *Titanic* ocean liner, 117 ff.
Rusticles, growth rates, 123
Rusticles, in water wells, 130
Rusticles, injection/extraction wells remediation sites, 131
Rusticles, lab evaluation of corrosive processes (illus.), 124, 131
Rusticles, relevance to maritime industries, 126
Rusticles, underwater pipeline problems, 134

Sulfate-reducing bacteria, rusticle component, 135

TCDD, tissue residues aquatic organisms, 22
Teratogenicity, azaarenes, 66
Tissue residues, contaminants in aquatic organisms, 1 ff.
Titanic, 1985 discovery of shattered hull, 117
Titanic, microbial rusticle growth, 117 ff.
Titanic, rusticle biomass, 128
Toxaphene, chiral insecticide, 91
Toxic effects, azaarenes range to different species (graph), 51
Toxic waste, steel drum deterioration on ocean floor, 134
Toxicity, azaarenes, 39 ff.
Toxicity, photoenhanced aquatic contaminants, 13
trans-Chlordane, CAS number, 107
trans-Chlordane, chiral insecticide, 90, 98, 104
trans-Chlordane, enantiomer structures, 92
trans-Chlordane, IUPAC chemical name, 107

INFORMATION FOR AUTHORS
Reviews of Environmental Contamination and Toxicology
Edited by
George W. Ware
Published by
Springer-Verlag New York · Berlin · Heidelberg · Barcelona · Hong Kong
London · Milan · Paris · Singapore · Tokyo

The original copy and one good photocopy of the manuscript, and a diskette with the electronic files for the manuscript, complete with figures and tables, are required. Manuscripts will be published in the order in which they are received, reviewed, and accepted. They should be sent to the Editor.

> Dr. George W. Ware
> 5794 E. Camino del Celador
> Tucson, Arizona 85750
> Telephone and FAX: (520) 299-3735
> Email: *gware7@aol.com*

1. Manuscript: The manuscript, in English, should be typewritten, double-spaced throughout (including reference section), on one side of 8½ × 11-inch blank white paper, with at least one-inch margins. The first page should start with the title of the manuscript, name(s) of author(s), with the author affiliation(s) as first-page starred footnotes, and "Contents" section. Pages should be numbered consecutively in arabic numerals, including those bearing figures and tables only. In titles, in-text outline headings and subheadings, figure legends, and table headings only the initial word, proper names, and universally capitalized words should be capitalized.

Footnotes should be inserted in text and numbered consecutively in the text using arabic numerals.

Tables should be typed on separate sheets and numbered consecutively within the text in *arabic numerals;* they should bear a descriptive heading, in lower case, which is underscored with one line and starts after the word "Table" and the appropriate arabic numeral; *footnotes in tables* should be designated consecutively within a table by the lower-case alphabet. *Figures* (including photos, graphs, and line drawings) should be numbered consecutively within the text in arabic numerals; each figure should be affixed to a separate page bearing a legend (below the figure) in lower case starting with the term "Fig." and a number.

To facilitate production, authors are strongly encouraged to submit their manuscripts (including figures and tables) in electronic form on diskette. Manuscripts may be submitted in DOS, Windows, or Macintosh format (but not UNIX) using popular word processing software (e.g., WordPerfect, Microsoft Word) or they can be saved as ASCII files. Tables can be prepared likewise or can be submitted as spreadsheets (e.g., Microsoft Excel). Figures may also be submitted electronically using such programs as Adobe Illustrator and Adobe Photoshop. Files can be saved both in .tif or .eps format only. Figures should be saved in both their original application and as PostScript files. Please provide tables and figures in their own files, not to be included in text files. Authors with questions regarding electronic preparation of their manuscripts are encouraged to contact Jenny Wolkowicki at Springer-Verlag via phone (212-460-1732), FAX (212-533-5977), or Internet (*JENNYW@SPRINGER-NY.COM*).

2. Summary: A concise but informative summary (double-spaced) must conclude the text; it should summarize the significant content and major conclusions presented. It must not be longer than two 8½ × 11-inch pages. As a summary, it should be more informative than the usual abstract.

3. References: All papers, books, and other works cited in the text must be included in a "References" section (*also double-spaced*) at the end of the manuscript. If comprehensive papers on the same subject have been published, they should be cited when the bibliographic citations extend farther back than to these papers.

All papers cited in the text should be given parentheses and alphabetically when more than one reference is cited at a time, e.g. (Coats and Smith 1993; Holcombe et al. 1995; Stratton 1999), except when the author is mentioned, as for example, "and the study of Roberts and Stoydin (1991)." References to unpublished works should be kept to a minimum and mentioned only in the text itself in parentheses. References to published works are given at the end of the text in alphabetical order under the first author's name and chronologically, citing all authors (surnames followed by initials throughout; do not use "and") according to the following examples:

Periodicals: Name(s), initials, year of publication in parentheses, full article title, journal title as abbreviated in the "ACS Style Guide: A Manual for Authors and Editors" of the American Chemical Society, volume number, colon, first and last page numbers. Example:

Leistra MT (1990) Distribution of 1,3-dichloropropene over the phases in soil. J Agric Food Chem 18:1124–1126.

Books: Name(s), initials, year of publication in parentheses, full title, edition, volume number, name of publisher, place of publication, first and last page numbers. Example:

Gosselin R, Hodge H, Smith R, Gleason M (1986) Clinical Toxicology of Commercial Products, 4th Ed. Wilkins-Williams, Baltimore, MD pp 119–121.

Work in an edited collection: Name(s), initials, year of publication in parentheses, full title. In: name(s) and initial(s) of editor(s), the abbreviation ed(s) in parentheses, name of publisher, place of publication, first and last page numbers. Example:

Metcalf RL (1978) Fumigants. In: White-Stevens J (ed) Pesticides in the environment. Marcel Dekker, New York, pp 120–130.

Abbreviations

A	acre	min	minute(s)
bp	boiling point	M	molar
cal	calorie	mon	month(s)
cm	centimeter(s)	ng	nanogram(s)
d	day(s)	nm	nanometer(s)(millimicron)
ft	foot (feet)	N	normal
gal	gallon(s)	no.	number(s)
g	gram(s)	od	outside diameter
ha	hectare(s)	oz	ounce(s)
hr	hour(s)	ppb	parts per billion (μg/kg)
in.	inch(es)	ppm	parts per million (mg/kg)
id	inside diameter	ppt	parts per trillion (ng/kg)
kg	kilogram(s)	pg	picogram(s)
L	liter(s)	lb	pound(s)
mp	melting point	psi	pounds per square inch
m	meter(s)	rpm	revolutions per minute
m^3	cubic meter(s)	sec	second(s)
μg	microgram(s)	sp gr	specific gravity
μL	microliter(s)	sq	square (as in "square m")
μm	micrometer(s)	vs	versus
mg	milligram(s)	wk	week(s)
mL	milliliter(s)	wt	weight
mm	millimeter(s)	yr	year(s)
mM	millimolar		

Numbers: All numbers used with abbreviations and fractions or decimals are arabic numerals. Otherwise, numbers below ten are to be written out. Numerals should be used for a series (e.g., "0.5, 1, 5, 10, and 20 days"), for pH values, and for temperatures. When a sentence begins with a number, write it out.

Symbols: Special symbols (e.g., Greek letters) must be identified in the margin, e.g. $A=\beta b/2\lambda$ [beta] [lambda]

Percent should be% in text, figures, and tables.

Style and format: The following examples illustrate the style and format to be followed (except for abandonment of periods with abbreviations):

Sklarew DS, Girvin DC (1986) Attenuation of polychlorinated biphenyls in soils. Rev Environ Contam Toxicol 98:1–41.

Yang RHS (1986) The toxicology of methyl ethyl ketone. Residue Rev 97:19–35.

References by the same author(s) are arranged chronologically. If more than one reference by the same author(s) published in the same year is cited, use a, b, c after year of publication in both text and reference list.

4. Illustrations: Illustrations may be included only when indispensable for the comprehension of text. They should not be used in place of concise explanations in text. Schematic line drawings must be drawn carefully. For other illustrations, clearly defined black-and-white glossy photos are required. Should darts (arrows) or letters be required on a photo or other type of illustration, they should be marked neatly with a soft pencil on a duplicate copy or on an overlay, with the end of each dart indicated by a fine pinprick; darts and lettering will be transferred to the illustrations by the publisher.

Photos should not be less than 5x7 inches in size. Alterations of photos in page proof stage are not permitted. *Each photo or other illustration should be marked on the back, distinctly but lightly, with a soft pencil, with first author's name, figure number, manuscript page number, and the side that is the top.*

If illustrations from published books or periodicals are used, the exact source of each should be included in the figure legend; if these "borrowed" illustrations are copyrighted by others, permission of the copyright holder to reproduce the illustrations must be secured by the author. Permissions forms are available from the Editor and upon completion by the original publisher should be returned to the Editor.

5. Chemical Nomenclature: All pesticides and other subject-matter chemicals should be identified according to *Chemical Abstracts*, with the full chemical name in text in parentheses or brackets the first time a common or trade name is used. *If many such names are used, a table of the names, their precise chemical designations, and their* Chemical Abstract Numbers (CAS) *should be included as the last table in the manuscript, with a numbered footnote reference to this fact on the first text page of the manuscript.*

6. Miscellaneous: *Abbreviations:* Common units of measurement and other commonly abbreviated terms and designations should be abbreviated as listed below; if any others are used often in a manuscript, they should be written out the first time used, followed by the normal and acceptable abbreviation in parentheses [e.g., Acceptable Daily Intake (ADI), Angstrom (Å), picogram (pg)]. Except for inch (in.) and number (no., when followed by a numeral), abbreviations are used without periods. Temperatures should be reported as "°C" or "°F" (e.g., mp 41° to 43°C). Because the metric system is the international standard, when pounds (lb) and gallons (gal) are used, the metric equivalent should follow in parentheses.

7. Proofreading scheme: The senior author must return the Master Set of page proofs to Springer-Verlag within one week of receipt. Author corrections should be clearly indicated on the proofs with ink, and in conformity with the standard "Proofreader's Marks" accompanying each set of proofs. In correcting proofs, new or changed words or phrases should be carefully and legibly handprinted (not handwritten) in the margins.

8. Offprints: Senior authors receive 30 complimentary offprints of a published paper. Additional offprints may be ordered from the publisher at the time the principal author receives the proofs. Order forms for additional offprints will be sent to the senior author along with the page proofs.

9. Page charges: There are no page charges, regardless of length of manuscript. However, the cost of alteration (other than corrections of typesetting errors) attributable to authors' changes in the page proof, in excess of 10% of the original composition cost, will be charged to the authors.

If there are further questions, see any volume of *Reviews of Environmental Contamination and Toxicology* or telephone the Editor (520–299-3735). Volume 159 is especially helpful for style and format.